Future Defence Challenges

Future Defence Challenges

Air Cmde (Dr) CN Ghosh

Manas Publications
(We Convert Fighters into Writers)
New Delhi-110002 (INDIA)

MANAS PUBLICATIONS
(We Convert Fighters into Writers)
(Publishers, Library Suppliers, Importers & Exporters)
4402/5-A, Ansari Road (Opp. HDFC Bank)
Darya Ganj, New Delhi-110 002 (India)
Off: ☎ 011-23260783, 23265523, Res: 011-23842660
Fax: 011-23272766
E-mail: manaspublications@gmail.com
manaspublications@yahoo.com
Website: www.manaspublications.in

Third Impression **2014 (FP-2003)**

ISBN: 13 Digit - 978-81-7049-158-3
10 Digit - 81-7049-158-4

₹ 795/-

Designed at
Manas Publications

Printed in India at
RK Print Services, Delhi
and published by Mrs Suman Lata for
Manas Publications, 4402/5-A,
Ansari Road, (Opp HDFC Bank),
Darya Ganj, New Delhi-110 002 (INDIA)

Preface

Revolution in Military Affairs (RMA) is the central idea to identify important trends in future warfare. Thomas Welch pointed out that, "RMA occurs when technological change makes possible the introduction of new material that when combined with organisational and operational change results in fundamental change in the conduct of warfare. What is important is not the speed with which a revolution takes place, but rather the magnitude of the change itself[1] ." Indian military would have to face these challenges of the future as demonstrated by the coalition forces during the Gulf war. Numerical superiority both in human and weapon strengths has given way to technological superiority and intelligence. Information (Info) technology and info-sphere have become another dimension of the future war not known before. Battle space is being redefined. Such critical advances will have an enormous impact on all military forces. The modern systems are sufficiently powerful to ensure that even their smaller number will have the ability to alter the course of the battle. That is why it is expected that wider access to this weaponry and associated technologies will increase the number of actors with sufficient military potential to upset the existing balances of power.

Indian Military has understood these challenges and are desperately trying to bridge the technological gaps. Space has finally found its place in the Indian Military doctrine. The Battlefield Damage Assessment (BDA) and vision elements are considered very actively. Info-sphere has become a new dimension to wage the 21st century war. Defence Research and Development Organisation (DRDO) have made some technological advances. But that won't be sufficient. The sensors systems are going through radical technological changes. Electromagnetic Pulse (EMP) and Directed Energy Weapons (DEW) would be applied in future conflicts. Non availability of precision guided munitions coupled with a non-responsive command control system wouldn't take Indian RMA very far.

Needless to say that RMA is also pushing the reinforcement of technical, organisational, and operational factors simultaneously across all fighting mediums—air, land, and sea and certain new areas of warfare are (or have been) emerging—like long-range and stand-off precision strikes, information warfare, dominating maneuver emphasizing the strategic positioning of forces, and space warfare.

However, if an adversary also has a weapon of mass destruction (WMD) capability, the stakes and the effects on strategy could substantially shift. WMD threats by radical, aggressive regimes may well force Indian Military into very different kinds of strategies. New modes of combat may have to be emphasized. Mobility, indirectness, dispersed, standoff, and disengaged operations may have to be included in the doctrines and exercised.

Post-Gulf war period brought in radical changes in military thinking. Indian Military would have to adapt itself to these changes and face the "Challenges of the Battlefield of the 21st Century," squarely. And to do that, they would have to address the following factors:

- Effects of parallel wars
- Value of precision weapons
- Importance of information at the strategic and operational levels
- Fragility of surface forces at the operational level

- Consequences of losing operational air superiority
- Dominance of air power
- Application of space

---

1 Thomas J Welch, *Revolution in Military Affairs : Our Perspective,* (Strength Through Cooperation : Military Forces in the Asia-Pacific Region), Institute for National Strategic Studies.

Contents

1

Military Power In 21st Century:

An Analysis

It seems that the first casualty of the 21st first century has been the most powerful and invincible country, the mighty United States of America. In a surprised move on September 11, 2001, four civil airliners were hijacked by the terrorists and committed the biggest hara-kiri attack of the history. World Trade Centre, the pride of America, and Pentagon, the bastion of the most powerful military, were hit by these kamikaze attackers. The US mainland never faced such fearful punishment. Thousands of American lives were lost. Many military leaders vanished under the rubbles of Pentagon. Military power of the super power looked utterly insignificant on the face of such unprecedented violence. That is why it is felt that military power will have its severe limitations to face such unknown challenges where there is no well-defined enemy and no national boundary to guard. The United States prepares to wage a war against such an enemy. Super power is pitted against unknown forces. Even a win, if at all, could go unnoticed.

Graying of society, increasing alienation and decline of state may be some of the factors, which will affect the military power equation in this century. Such extraneous factors are not directly related to military yet it would have profound effect, when we seek to understand the balances of power in future. Growth of authoritarian regimes like in the previous century under whom military power grew unabated is not conceivable in this century, who earlier based their power-balancing act mostly on military factors and could not satisfy the society as a whole. Therefore, there is a requirement to analyse the shape of the military power of the 21st century.

Technology and information driven society will not be ready to absorb many impacts brought in by these ultra new factors. Military power and the military balance in the third world countries like India will be at crossroads, unless the changes are understood and absorbed quickly. It will be better to understand the Orwellian Big Brother factor, which is no more a figment of imagination. These are as true as it was described. No military proceedings will go unnoticed. Powerful sensors and satellite borne cameras beaming directly from space will broadcast everything that could be seen or sensed. This will create a unique situation for a military commander, who will have to think of the impact before he could fire his first gun. Such scenarios will raise many military related questions. Classical theories given by Clausewitz or Jomini will be questioned.

Further complication will rise due to growing interdependence and spread of market mentality. States will have to act in a particular way dictated by such forces, in preference to narrow sense of sovereignty. Dilution of national interest is perceptible, unthinkable in the previous centuries. Economic barriers will be smoothened at the cost of general power spectrum. European nation states will find it difficult to be guided by the norms set by them during the last two centuries. Lebensraum will not find many takers, despite such strong ethnic divide. Napoleon and Hitler will be a permanent part of history, not to be repeating in any foreseeable future. Military power structure has to be carefully balanced leading to a type much different than what was before, mainly guided by such growing interdependence in both economic and military fields.

Added to all these will be the need to understand the emergence of regional hegemonies. A subtle but shifting balance of power is taking place in its new reincarnation. Display of power is assuming a different route to create such hegemonies.

Where there will be no chance to complain but understood easily and accepted with not many alternatives left. Decay/ implosion of regimes may cause such regional hegemonistic growth plausible and mostly will be visible in the third world countries of Africa and Asia.

Against such a backdrop we find proliferation of anti-national, non-state actors. They are gaining ground steadily as against the state actors and can at times tilt the balance either way. Effective governance by the state is the precious casualty to this new type of actors. A non-reversible trend has been established.

21st century is going to face a typical challenge that did not exist before. Genetic engineering will quickly get appropriated by the military mind and deployed with the existing system to overpower the adversary and impose its will. Human genome programme has been well established. The scientific community is not very far away from creating a new paradigm, which was not known by mankind earlier. And such technological revolution will find its way into military application. Because, military power has an insatiable urge to absorb new technology. The ultimate desire will be to dominate and deny somebody, something or use the same for annihilation of whatever is good for the mankind.

Growing consciousness of environmental protection will be another force, which will not permit indiscriminate use of chemical, or biological weapons. Vietnamese jungles could not be destroyed any more like the way it was done during the Vietnam war. Military planners will have to be careful to take all these factors into consideration. Environmental destruction could be taken as a crime against humanity.

And finally while analyzing military power in the 21st century it would be necessary to understand the uni-polarity of the world, decline of the Soviet military power, Chinese desire to fill the vacuum. Decline of the role of United Nations in a conflict situation also needs to be understood. It is not clearly known or understood what would be the role of the United Nations, in the face of increasing competition in economic field and in cultural norms.

Conflict situation in the society will have its source as the struggle between haves and have-nots. Increasing alienation will give rise to terrorism. This ultra new factor of 21st century will challenge traditional military power and its basic constituents. And the idea would be to analyse and understand the constituents of military power and its effectiveness in the changing realities of the 21st century.

Sources of Military Power

Over the years different analysts in various ways have defined the power of the nations. Perhaps the most popular dictum regarding power is that of Mao Ze Dong, who said "Power flows from the barrel of a gun". This saying gained its popularity largely due to its pithiness and vivid imagery, rather than due to any profound content. Then again, power has been described as the ability of a nation to hold its own against other nations and also as ultimately, the ability of a nation to wage war. Perhaps one of the most acceptable formulations, which define power, is that "Power in international relations is the ability of a nation to use its tangible and intangible resources in such a way as to affect the behavior of other nations to its own advantage." Derivation of national power takes the following shape:

- Power is not absolute; it is relative to the power of other nations.
- Power may also be considered to be relative to place and time.
- It may be possible for a nation to exercise its power or influence on another nation effectively at a particular place and at a particular time. But this may not be possible at other places or at other times.
- Power is also dependent upon many conditional factors, depending upon the international environment. This conditionality of power is a factor of growing significance.

- In specific conditions, that is, within a region or between two states or between two grouping of states, power is finite or as it is popularly said, power is a "zero sum game". That is, between two states, as one grows in influence, it is logical that, relatively, the influence of the other wanes.

Following are the major factors that are to be taken into consideration to understand source of power:

- The first source of national power may be termed geographical. And India enjoys a unique situation to this respect. High barriers of Himalayas in the north and bounded by the seas all around gives it a peculiar strength. Experts in the field undoubtedly give the greatest weightage to the geographical or the geo-political factor. Whether this factor is equally important today is a matter of question. But it cannot be denied that essential features of a nation's geography, such as its location and access to the sea, still constitute key factors in determining its independence and its ability to exert influence on its neighbors and beyond.
- Population has always been termed as the biggest drag for the growth of a nation like India. But the strategists consider that national power has direct relationship with its population. That is why a country with a large population has direct resource potential within a country. But the size of a population is an advantage upto a point and it must be related to the abstract concept of quality of a nation's population, because a large, illiterate, impoverished and disgruntled population will in fact be a tremendous handicap to any nation in its search for power. That is why, India's greatest concern should be the development of human resource in this century.
- In whatever term we prefer to identify this source, it is clear that the quality of a nation's leadership, its responsiveness to the public good and sensitive to the aspirations of its population, its resoluteness in maintaining

its international interests, and its astuteness in managing its international relations to protect and advance these national interests are indeed a key source of national power. India prides itself to be the biggest democracy of the world but polity of the nation is desperately fragmented, as such national leadership is incapable of providing a clear set of goals for the nation.

- A country that has natural resources in abundance will be able to maintain a higher degree of independence in its relations and consequently, the power factor of such a country is enhanced.
- An abstract factor like National Image could be an added factor for military power. The national image that a country enjoys depends again on its perception by other countries, which is a factor that can be controlled by skillful use of diplomacy and projection of the nation's image in its foreign dealings.

Elements of National Power

There is an increasing realisation that military power has growing limitations and the emerging international order is likely to reduce the significance of military capability as an ingredient of national power even further. Yet, to understand the national power such an important ingredient cannot be totally set aside.

We have discussed the sources of power, which clearly indicates that national power has certain basic elements. Military power forms a very important segment, which is the most visible and easily understood element of national power. Modern military machine can be deployed rapidly against any adversary provided there is an asymmetry of military balance between two nation states. The will to use such machinery has an overpowering impact on an adversary. But world community is not ready to accept such display of crudity, which was clearly visible in the Stalin or Hitlerian era of the previous century. In fact, all power can be perceived as of little consequence if in the long run, it is not possible for a nation to enforce its will at the required time.

Next comes the economy factor. Economic capability is the key to a nation's development and well being. In today's world economic power is economic capability, and economic interests are the fundamental driving force behind international relations. It can be convincingly argued that most conflicts today will be due to economic causes rather than any other. Globalisation on the other hand makes many national security functions redundant. Cross-border integrated production and the nationality of a company is becoming increasingly meaningless. Moreover, the fight to secure physical resources, which dominated most of the nineteenth and twentieth centuries, is completely absent. But globalisation has its ill effects also and causes many other security related problems, which will become more and more acute over the years. Open world economy will make the state of the art military technology more easily available. The number of countries with nuclear, chemical, biological and laser weapons will rise. Vincent Cable describes these new phenomena as "The redundancy of traditional forms of national security has led to attempts to reinvent the language of national security for a world of economic globalisation[1]."

National power is also directly related to technological capability of a nation. Though technological development has direct relationship with economic development yet it is not possible for a nation to have economic advancement without technological capability. In reality, a combination of technological capability and economic power is essential for a nation to be able to build itself up or advance economically. There are preconceived notions by certain analysts that Indian atomic explosion and India's display of space capability have minimal effect on the country's security interest. Which is totally misfounded because Indian nation now has better standing among the technologically advanced nations. Neil Chamberlain mentioned that" The nations which can embark on military actions to improve their position in the international authority structure and in the distribution of the international advantage will of necessary require a substantial industrial base.[2]"

That national power flows out of ideology, is a very important factor that needs clarification. Political and religious issues have historically been the cause of many conflicts but a nation, which has ideological cohesion, is likely to be at an advantage compared to` another which is ideologically or religiously disparate and where ideology instead of being a unifying force, is a source of conflict and division within the nation. The significance or importance of ideology as an ingredient of national power can be assessed in other ways also. There is a section of opinion, which holds that ideology as a motive force in international relations is on the wane; and to substantiate this thesis, the break up of the Soviet nation can be cited as the best example. China being a communist country is rapidly moving towards market economy. Ethnic integration is emerging as the new force for combination. Finally it can be said that ideology is of diminishing significance in the ability of one nation to influence the actions of another- such as in the case of the communism-capitalism confrontation. However, it still has a role in generating national morale and in creating national cohesion and unity, and to this extent, it must be accorded its due place as an ingredient of nation's power.

Opinions of scholars may vary on the relative importance that should be accorded to each of these elements of national power. But it is very clear that without the prior development of an economic base, the build up of a military capability can lead to serious consequences within a nation. The USSR is the best-known casualty and so are many other smaller nations. It is however clear that under different circumstances each element of power as discussed above, assumes greater or lesser significance. And also there are those who would argue that the world order is exhibiting tendencies to create conditions where military power will become inapplicable or irrelevant altogether. This however is a difficult theory to subscribe based on the evidence before us today. Paul Kennedy in his book, *The Rise and Fall of the Great Power*[3], has said in reference to Japan: The history of the past five hundred years of international rivalry demonstrates that military security alone is never enough. If the nation overextends itself geographically and strategically, even if, at a less imperial level, it chooses to devote a large proportion of its income to protection, leaving less for productive investment, it is likely to find its economic output slowing down with dire implications for its long term capacity to maintain both its citizens, consumption, demand and its international position. This was not realised in the former Soviet Union.

Economics and military strength are important determinants of national power and these at various times assume predominance over each other. The 18th century economist Adam Smith[4] summed up by stating, "The first duty of a sovereign, that of protecting the society from the violence and invasion of other societies, can be performed only by means of a military force. But the expense of preparing this force in times of peace and of employing it in times of war is very different in different states of society in different periods of improvements."

Malthus[5] believed that economic power and military power enjoyed a kind of cyclic predominance. He believed that as the economic power enjoyed a kind of cyclic predominance, the economics of a state was built up, it developed its military power to protect its own wealth and since war itself paradoxically often results in prosperity and the accumulation of wealth. Thus there is a cyclical relationship between economic wealth and military power. Hence, this was used to explain the periods of prosperity and war that were seen in Europe towards the eighteenth and nineteenth centuries. But can that factor be applied in the present or future context of this century? On the other hand Indian geopolitical situation will never correspond to such scenario as described. Being the largest nation in the region India will never be in a position of negative asymmetry in relation to its neighbours considering the present trend.

Judging by today's trends, economic factors are likely to become predominant in international relations. However, the image of a nation does depend to a great extent on its military capability and which is not likely to conveniently disappear from the anatomy of nation's power. For example, the attitude of the Western powers and of the former Soviet Union to China was largely based on that nation's military capability because China was not an economic power a few years back. But such a state has totally changed now due to the fall of Soviet empire and the emergence of China as an economic giant. It must also be accepted that an unhealthy emphasis on building up the military capability at the expense of economic development can lead to disastrous consequences for a nation in its developmental process. Indian military planners must learn from the predicament of the Warsaw Pact countries.

Equation to Measure Military Power

Ray Cline[6] has postulated a means of calculating the relative power of nation states. He developed the following formula:

$$P = (C+E+M) \times (S + W)$$

Where C is the critical means of nation and consists of two components - its population and territory. E is the economic might of a nation and is also sub-divided into several factors, which are basically parameters or economic indicators such as consumption of steel, GNP, foreign trade, balance of payments, food production and energy production. These are summated into a single figure, which, however hypothetical, is still a relatively reliable indicator of a nation's economic standing. M is the country's military capability and here a very clear distinction has been made between nuclear and non-nuclear military capability. Notwithstanding all the moralizing that any developing country, particularly India, is subjected to by the international community, it must be understood that post-1998 nuclear explosion the global view of India has changed. India is in a unique situation of having two nuclear powered neighbours which no other nation can boast of. Possession of nuclear destructive capability makes a great deal of difference to the way a particular nation is perceived by other nation states. In conventional military might Cline has used as his basic indicator the size of the defence budget as a percentage of GNP. This may be questioned as it is not the size but the quality of an army that really indicates its military might. The sum of these three factors—critical mass, economic might and military capability—is multiplied by Cline by a combination of a rather abstract quantity, which he calls national strategy (S) and national will (W), which together represents a nation's political will and leadership. This, naturally, is a factor, which is less than one, and therefore serves to reduce the total sum of a nation's economic and military might by this particular factor. Whereas national strategy must take into factors like ideology of a country and the clarity of its national purpose. And in national will would be included such factors as the quality of its leadership, resoluteness, its responsiveness to its population's demands and pressures. Together these two factors constitute that element which is in the realm of the abstract, that is the socio-political factor. Whereas one can agree with Cline's principle for calculating power, the actual manner in which this has been quantified is open to question because rating a state in such a manner will be erroneous. This can be a merely a matter of subjective assessment. Cline's calculus is therefore of great benefit in enabling some kind of relative assessment to be made of specific states and nations as long as a very clear understanding is first reached about the conditionalities under which such assessments are made. Such an exercise is sometimes necessary for purpose of evaluation and comparative assessment. But it must be borne in mind that elements of power, like the steel production or the size of armies, should themselves never be equated with power itself.

Nuclear Weapons

It is considered as a source of power for a state which could harness nuclear energy and convert into a weapon. But it may be appreciated that political impact of the nuclear weapons has been very little because no one has come up with any idea as to how a nuclear war could be fought. The super powers also found out that nuclear weapons do not confer great advantages in their dealing with other states without nuclear capability. The military strategists are yet to figure out a way to carry out any conventional war in the wake of a nuclear attack and implement a type of force structuring. Decline of conventional forces of the super power was clearly visible in the cold war era but they retained a highly sophisticated conventional capability. Which is not visible in the case of India and Pakistan even though they have displayed their nuclear might. Martin Van Creveld stated that, " The principle reasons why military powers for many years devoted so much effort preparing for conventional conflict in a nuclear age was the imperative desire to prevent nuclear war from breaking out[7] ."

The international power environment obtaining at that time will largely condition the emergence of any state as a major power. Before analysing any country's potential and capabilities as a power, the current trends in the international power structure must first be understood and an assessment made of the pattern of power relation that is likely to obtain in the near future. It may be argued very well here that such factors do not fare well with the study of the analysis of military power in the 21st century. But if carefully examined it may be seen that all the factors mentioned above will cast their shadow on such a study. As the indirect effects are not clearly visible they should be read, understood and their implications have to be taken into consideration more than the visible factors. "So in the most part of the world no man, woman, and child alive today will be spared the consequences of the newly emerging forms of war[8] ."

Complication will also arise due to inter-relationship between globalisation and national security because all the states will not be at the similar level of development. Globalisation is fairly advanced in many countries of the Western nations and it is an irreversible fact. This could become a major constraint for those nation states. There would be a tendency to relate everything in terms of the traditional balance of power, in countries like India who have recent history of conflict with its neighbours; and integration with the outside world will be limited. Third group of countries will be the nation states of Africa where national identity is weak due to factors like poverty, and general lack of integration with the outside world. Therefore military power will be totally subservient to this reality of the future and will be totally dependent on the non-traditional way of functioning of the nation states.

It should be pointed out that during last decade major conventional wars between states have been few. The confrontation of Iraqis with the US led qualition forces and the Serbian resistance against NATO forces cannot be called classical battles fought between the nation states in the last century. On the contrary countries seem to be on the brink of civil war and face occasional outbreaks of terrorism that cannot be identified with any country. And that is why such conflicts cannot be termed as Trinitarian as written by Clausewitz. There is a clear paradigm shift. Trinitarian wars are being superseded by non-Trinitarian wars and most of the nations including India are at a loss to deal with it effectively. Success against them has not been very clearly defined. The military power and its role in this century will be very much different from its defined role as understood by the mankind. That is why the main thrust of Clausewitz philosophy of war is questioned, when he states that " states wage war using armies in pursuit of political objectives." Post 11 September 2001 developments in the world may be a clear indication that Clausewitzan theories may not be applicable to such new types of war where there are no well defined borders. State is not the main actor and the enemy is unknown whose strength and weaknesses cannot be assessed with any amount of surety and the principles applied in classical sense.

Notes :

1 Vincent Cable, *Globalization and Global Governance*, 1999, The Royal Institute of International Affairs. pp40-45.

2 Chamberlain Neil W Neil, *Beyond Malthus: Population and Power*, 1970, Basic Books, Inc New York. p182.

[3] Paul Kennedy, *The Rise and Fall of Great Power*, 1987, Random House, New York.

[4] EK Hunt and Howard J Sherman, *Economics: An Introduction to Traditional and Radical Views*, 1972, Haper Row Publishers,New York. pp33-34. "Publication of Adam Smith's *The Wealth of Nations*, a new individualistic philosophy – classical liberalism – had definitely gained the ascendancy in England. This individualistic philosophy had existed throughout the mercantilist period.... In the end the new classical liberalism prevailed because it reflected the needs of the new capitalist order."

[5] EK Hunt and Howard J Sherman, Economics : *An Introduction to Traditional and Radical Views*, 1972, Haper Row Publishers, New York. pp 49-50. "Thomas Robert Malthus' population theory was an important and integral part of classical liberal economic and social doctrines."
Chamberlain Neil W Neil, *Beyond Malthus: Population and Power*, 1970, Basic Books, Inc, New York. p3. "In his first chapter of his Essay, Malthus asserts that " I think that food is necessary to the existence of man. Secondly, that the passion between the sexes is necessary and will remain nearly in its present state. Assuming then my postulates as granted, I say, that the power of population is indefinitely greater than the power in the earth to produce subsistence from man. Population when unchecked increases in geometrical ratio. Subsistence increases only in an arithmetical ratio. A slight acquaintance with numbers will show the immensity of the first power in comparison of the second. By that law of our nature which makes food necessary to the life of man, the effects of these two unequal powers must be kept equal."

[6] Ray Cline, *World Power Assessment: A Calculus of Strategic* (Boulder Cdo; West View Press 1975) as quoted in *Power Strategy and Security*, ed Klans Knorr, (Princeton University, 1983) p.14.

[7] Martin Van Creveld, *On Future War*, 1991,Brasseys, UK. p5.

[8] Ibid, p225.

2

Revolution In Indian Military Affairs

There is no standard definition of Revolution in Military Affairs (RMA). Many military journals have discussed the RMA and its impact on the future wars, which will be characterised by speed, intelligence and precision. No doubt, the older form of warfare will co-exist with the newer form driven by RMA. This newer form will differ from its antecedents. Soldiers with higher IQs will conduct future forms of wars; sophistication in weapon system will invite better quality of decision cycle. Greatest need would be to act within the enemy's decision cycle. That is why Institute of National Strategic Studies stated, "A final feature of a definition of RMA is a 'culminating event', a battle that employs new systems, operational concepts and organizations and that clearly demonstrates a dramatic change in the conduct of warfare[1] ".

We do not have the correct tools to assess RMA. Neither do we have any good model to emulate in Indian context. There are certain concepts of operations and military capabilities (present and future), which are not sufficient to understand such transient nature of the future war. And it is certain that the Indian military will face the wars in 21st century characterised by:

- Greater availability of information, hence greater transparency in all facets of battle.

- Duration of the war will decrease but it will be marked by higher tempo of activities than before.
- Increased accuracy, lethality and stand off capability of weapon delivery leading to enhanced reliance on mechanised assaults and missile strikes.
- Increased mobility of opposing forces, leading to fast changing operation and tactical situations. The classic sequential progression of battle graduating from units and formations to theatre commanders may no longer hold good. In other words, the battlefield would become non-linear. This would call for speedy decision making, greater delegation of authority for freedom of action to subordinate commanders.
- Increased reach of integral firepower and surveillance resources, including space based and underwater systems, would enhance the battle space as described earlier.
- Teeth to tail ratio for the armed forces to be much tighter.
- Collateral damages to be kept to minimum, considering the enhanced role of the media.

Technology and information driven warfare of the future will strive to remain inside and ahead of the enemy's information processing and decision cycle. Considering that, the chapter discusses the necessity to understand the Revolution in Military Affairs (RMA) as a concept, and examines its applicability to Indian military.

An American scholar Michael Vickers[2] has identified a number of cases, which satisfy the definition of RMA.

- "Those nations that became aware of the importance of technological change and adapted operationally and organizationally accrued major advantages. Others then had to try to catch up rapidly. An example is the German Blitzkrieg and the Allied attempt to master this new approach to armoured warfare.

- "New, very different measures of merit, of effectiveness, became relevant, the old ones are more quickly discarded by the winners, who learn to adopt the new ones.
- "Some material organizations and operational concepts decrease in importance (sunset system) in these periods while others increase (sunrise systems). Winners are sensitive to these changes and act on them.
- "Losers planned for the future by using their current organizations and practices as a point of departure. They were content to graft technological change onto existing structures and operations and to improve them at the margins."

Revolution in Military Affairs has become a common terminology. Many military strategists are trying to explain this new terminology from their point of view. But what would have Clausewitz, the master military theoretician, would have said? Clausewitz clearly pointed out that the states wage wars using armies in pursuit of political objectives. Is it valid on the face of this new thought like RMA?

Some writers are maintaining that nuclear weapons, transnational warfare, counter-terrorism, counter-narco trafficking and increased compartmentalisation of political and military leadership in this generation have rendered Clausewitz's definition of war obsolete. This is also said that future wars will be fought not to pursue national interests but to kill enemy leaders. There are necessities here in India to explore such issues like impact of RMA on the structure and philosophy of the 21st century armed forces.

However, it must be understood that while older form of warfare will continue to co-exist with newer ones, the RMA when it is complete or if at all complete, will mean that the conduct of future war will include soldiers with higher IQ, knowledge oriented weapons, and five dimensional warfield that includes, breadth, height, depth, space and time. Time will be the most critical factor in this transformed scenario. Sensor to shooter time loop has to be condensed further. Ability and need to act within the enemy's decision cycle will be the requirement. Capability to attack simultaneously, and with precision at the tactical, strategic levels. It is not only military that is going to get affected by such ideas; the RMA is going to pose serious challenges to decision-making apparatus of a state as an institution.

That is why the question is being asked, what is the relevance of Carl Von Clausewitz today, who developed his theory more than one hundred and seventy years ago in Germany? Someone goes to state that, his masterwork *On War*[3] has outlived its philosophy due to such rapid changes of technology and changes of the complete world order brought about by economic imperatives. Interplay of factors like rapid technological advancements in weapon systems, information revolution, increasing interdependence, decline of the state, proliferation of transnational non state actors, spread of market economy etc., can erode much of the writing by Clausewitz. Contrary to this, there is another school of thought, which completely supports Clausewitz and negates the view expressed earlier. The biggest question arises here, how to match Clausewitz with RMA when he never mentioned anything about the same in his writing.

The battle space is nothing but physical volume in which the military forces carry out their operations and that includes infosphere as well. And Indian battle space is undergoing certain changes. That has to be examined to understand Revolution in Indian Military Affairs. Operational, organisational and material changes resulted for the most part from each Service's efforts to enhance its efficiency. But it is now well established that due to rapid technological advancements, the battle space is getting redefined. The sensors and weapon systems are reaching well beyond a Service's sphere of influence. This may be called as "overlapping battle space" resulted by Revolution in Military Affairs. And information technology would remain as its epicenter.

Battle Space Information System[4]

That is why operational concept of the battle space information system needs to be understood.

Once the battle damage assessment has been carried out, operational appraisal is done. Then follows the operational objective definition, sensor tasking management, information acquisition, situation assessment, and finally the resources are allocated.

From here the operation planning begins till such time the strike is handed over to the shooter. The concept as defined is explained in the following diagram:

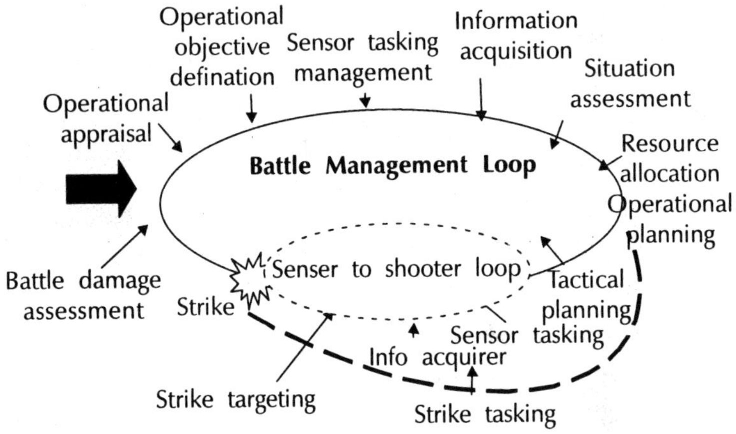

Diagram 1. Battle Management Loop is the essence of RMA.

These may be the essence of entire revolution of the military operational concept of this century driven entirely by information technology.

Information Warfare (IW)

The entire system discussed above and its success depends on information collection, its processing and application. Information technology is the key for modernisation and growth towards self-reliance and wage war in infosphere. On the other hand, it has opened up the much-dreaded Pandora's box. Such type of warfare can disrupt military command and control system and cause serious economic damages. It may be understood that any interjection in the loop described in the diagram, could have serious repercussions. India is yet to attain such status similar to the advanced nations, but is gradually creeping towards that eventuality where sensor to shooter loop is getting tighter and vulnerability to hacker attacks becoming a reality.

Indian military must understand its importance, and take this into account with its relevance for RMA of this century. Key technologies to make a thrust in this field will be possible by a right mix and integration and knowledge in a meaningful time frame. Technologies those could be used effectively are:

- Communications and networking
- Automated fusion of imageries
- Automated target recognition
- Satellite technology for communication, communication intelligence (COMINT), electronic intelligence (ELINT)
- Global positioning system (GPS) and mapping
- Sensors:
    - o High efficiency radio frequency source
    - o High energy lasers

There are certain special characteristics of the Information Warfare. These are:

- **No International Borders:** Like other forms of attack, Information related attacks could be carried out from anywhere in the world which do not respect the nationality or national boundaries. There are no ways to know who is under attack and who is attacking, and what is the difference between trans-national crime and warfare.
- **Low Cost Warfare:** Adversaries can acquire the necessary equipment to wage this type of warfare at a very low cost. They would require a few computers and good connectivity to attack any network or steal information from them. Even if the entry is restricted to network,

determined attacker will find a way to intrude and do the necessary damage. But the question arises, whether such an intrusion could change the course of the future battle. Martin Libcki, the main exponent of Information warfare, has clarified that such warfare if carried out against command and control system could disrupt military operations.

- **Information a Tool of Subversion:** Use of Internet to build up political support or work on the psychological factor of a nation has been spoken about most glibly by the advanced nations. But in a country like India, it would not have such catastrophic effect as has been portrayed by the strategists. Jammu and Kashmir Liberation Front (JKLF) and many other terrorists' organisations have been maintaining immaculate websites for a long time, which has not created the expected psychological impact either nationally or internationally.
- **Tactical Warning and Attack Assessment**: Nature of threat is rapidly changing and the targets are also not as clearly defined as it used to be. As a result, traditional methods of intelligence gathering and analysis would be taking a back seat in the face of the challenges thrown up by the IW environment.

Limitations of information warfare has to be reviewed at a different plane in the Indian context, despite the doomsday effect forecast by its proponents. The experts in this field are not convinced about its invincibility or its ability to disrupt or defeat a country soundly to conquer a nation. Even Prof Libcki pointed out that " In its ability to bring a country to its knees, hacker warfare is a pale shadow of economic warfare, itself of limited value. Suppose that hackers could shut down all phone service (and, with that, say, credit card purchases) nationwide for a week. The event would be disruptive certainly and costly (more so every year), but probably less disruptive than certain natural events, such as snow, flood, fire, or earthquake – indeed, far less so in terms of lost output than a modest size recession[5]." This may be fairly optimistic but it can be stated definitely that information warfare couldn't be as destructive as other weapons of mass destruction. That is why, it should be understood and analysed, whether IW system could be as effective in imposing a nation's will as traditional firepower. If so, then it should be included in the list of weapons and may have certain legal aspect for its use on an adversary.

RMA: Indian Perspective

Indian scenario is not very comfortable, because as it stands, the Indian military will be way behind the technology driven wars of the 21st century. With Indian Military planning, bereft of cohesiveness and prone to knee jerk reaction, procurement would not match emerging technologies and will gradually lose the race in the advancing technology of the future, if radical measures are not initiated.

On the other hand, the Defence Industry has remained where it was decades ago, with overstaffed ordnance factories barely managing to run with Government subsidy, notwithstanding a captive buyer like the Indian Military. As such, they will find it hard to match the Indian Military requirements of the 21st century. The availability of cutting edge technology to the armed forces would be difficult. And it is certain that technologically advanced nations would not oblige India to the effect. Broadly speaking, Indian RMA will depend on the following:

- Strategic systems, which are essential for the country's defence, but not available to us.
- Systems and Technologies, which leap frog into bridging the gap with technologies likely to be developed in the future.
- Integration of the command and control network of the three Services.
- Synergy between the Services and Defence Research and Development Organisation (DRDO).

- Effective utilisation of the technologies available in the private sector.

Indian Military has embraced the indigenous scientific and technological developments in various fields, to be incorporated in their operational planning. These have been discussed in the successive paragraphs.

Remote Sensing Satellites: Indian remote sensing satellites may be among the best that are circling the earth at present. The imagery is much better than the SPOT and other Landsat group of satellites. As of late 2000, five IRS satellites were in the orbit and more are scheduled to join this family. The Indian Space Research Organisation (ISRO) and its commercial arm Antrix Corpn Ltd successfully launched the IRS-1D the Earth Imaging Satellite on 29 September, 1997, from Sriharikota, which is similar to IRS-1C. These dual use satellites provide 5.8 mtr resolution images to customers, twice as often as was possible with just the IRS-1C. This would benefit the three services immensely.. This resolution, which as of early 1998, was the best among the civilian remote sensing satellites in the world and is superior to the 8 mtr resolution initially reported for the panchromatic imager[6].

IRS-P4 (OCEANSAT-1)[7]: Following the above, IRS-P4 was launched with payload of Ocean Colour Monitor (OCM) device which is a solid state camera operating in push broom scanning mode, using linear array Charged Coupled Devices as detectors. This camera has 8 narrow spectral bands, operating in visible and near infrared bands with a spatial resolution of 360 mtrs, and swath at 1420 km. As ISRO explains, since OCM is designed to provide the highest spatial resolution among the contemporary ocean colour sensors with unique regional and global coverage, it is expected to provide ample opportunities/ application potentials, for both scientific, commercial and military.

Imagery Intelligence (IMINT) and Military Application Satellites: The satellite imagery, battle damage assessment etc., helped military planning immensely on day to day basis in both Gulf and Kosovo wars. And that may have prompted the Defence

Services to come to an understanding with ISRO and plan future defence application satellites.

Unmanned Combat Aerial Vehicles (UCAV)[8]: Experience of the Gulf and Kosovo wars brought several strategies to the forefront, and the supremacy of air power has been clearly established. It has been established that air power, however, made it easy for the ground forces to complete the annihilation of the Iraqi Army. That is why the war strategists do not have any doubt that some form of air power will decide the course of the battles in future. Such deliberations naturally bring the factor of Unmanned Combat Aerial Vehicle (UCAV) in the forefront to augment air power. UCAV has not only the attraction of keeping combat casualties low but overcomes many human limitations. UAVs will be smaller, faster and more agile than manned fighters. Such unmanned vehicles could be used in preference to manned flights in a dense air defence battle space optimised by low looking radars and QRMs (quick reaction missiles). These will be very useful to keep a long vigil on the enemy activities and pass information to the command post. UCAV concept is technically feasible. These can meet the operational need and fit into a sound plan of operations. Historical biases for manned aircraft should not be the stumbling block for embracing this technology in the futuristic technology driven wars. These forces need to be understood and application made accordingly. And this was adequately proved during the recently concluded Afghan war. The Predators of US Air Force carried Hellfire Anti-tank missiles and were controlled by ground crew effectively. Following the success of the Predators, the Global Hawks were pressed into service for surveillance and communication in Afghanistan. Indigenous UAVs are under development by DRDO. And before these are made available to the Services, Indian Air Force has taken a step in the right direction by inducting UAVs like Searcher II and Herons to its inventory through procurement route.

Defence R&D: Defence R&D is being handled entirely by Defence Research and Development Organisation (DRDO). This organisation will have to work very hard to meet the future requirements of the armed forces in this century. A viable match between DRDO capabilities and the demand of the armed forces should emerge to resolve the demand/supply crisis. Mutual understanding, penalty driven MoUs, user/DRDO interface, would be the requirement of the future.

Time is running out for Indian Military to step into 21st century. The R&D in the core sector has to be given an impetus. Increment of the DRDO budget is not the only answer. Financial assistance, in terms of MoUs, needs to be formulated with the DRDO and with the civil industry as well. India is neither short of manpower or brainpower. It lacks direction.

DRDO[9] has grown over a period of last 40 years or so and handles the entire defence related R&D. Self reliance if at all could be with the help of scientists of this organisation. Some of the important areas where DRDO has excelled are given below:

- **Millimeter Wave System:** MM wave technology has been the intellectual property right of the advanced nations. DRDO's efforts to harness this technology will assist guidance and control of airborne military sensors, through large gain channel bandwidths. This field will be critical to force multiplier systems.
- **Radar:** Phased Array radar, apart from others, which has been developed by DRDO, should be in consonance with the present generation requirement of the armed forces. This radar has significant advantages over other control and guidance radars.
- **VLSI and Microwave Monolithic Integrated Circuits (MMIC):** A commendable achievement by DRDO was to develop MMIC. Another critical technological breakthrough, which would help development of EW equipment, is Very Large Scale Integration (VLSI).
- **Lasers:** Lasers have assumed great importance because of their wide ranging applications. DRDO is in the process of developing high power laser weaponry and other laser related technological innovations.

- **Missiles:** The Integrated Guided Missile Development Programme (IGMDP) has been a success story. Prithvi has already entered the service with both Army and Air Force. Agni is also being inducted. Trishul and Akaash should be in the inventory soon. Along with this Nag is also in a very advanced stage of development.
- **Aviation**
  - o The DRDO-Air Force collaborative effort to fly Light Combat Aircraft (LCA) has seen the first prototype in the air this year. And with the American sanction being removed the aircraft should receive the right push to get into squadron service.
  - o Similarly the Advanced Light Helicopter (ALH) should be able to reach the production stage shortly.
  - o Remotely Piloted Vehicle (RPV). This project is also in a very advanced stage of development. It should soon be included in the active service.
- **Core Materials:** The right steps have been taken to produce leading edge material like Al-lithium alloy, magnesium-lithium alloy, ultra high strength steel, titanium alloys, carbon ceramics, and ceramic composites, etc. However, cost and commercial viability will need to be studied and if necessary start up subsidies provided.

Requirements of Indian Military in 21st Century

Indigenous scientific achievements, no doubt, would assist Services to incorporate them in their operational planning and force structuring. But the following developments must be taken into consideration by Indian Military for them to achieve RMA in real sense.

Sensors: Sensors are nothing but physical entities, which convey information about external environment from the sensing domain by using hardware and software. These are designed to interact with the sensing domain and with the user or machines to which the sensor provides information and schemes for producing, storing and analysing the information that the sensors gather. Advances in sensor technology are increasingly creating opportunities to coordinate the operations of individual sensors in ways that produce significant operational benefits. This needs a fundamental and comprehensive examination of sensor requirement and their deployment. Constant development of the system keeping the cost factor into consideration will be essential.

The growing need for sensor derived information, coupled with limited number of systems in an affordable force structure, creates ever growing demands on the performance of sensors of all types. Sensor technology has been making rapid strides. There are certain sensors which will make their appearances in this century and have profound effect on military affairs.

Electro Magnetic Pulse Bombs[10]: During the Kosovo war a Tass wire story reported for the first time about the presence of a new weapon being tested by the USA. This was the most secret weapon designed to destroy radio electronic equipment. The Russians also went on to report that these bombs were dropped by B-2 stealth bombers to generate electromagnetic pulse (EMP). Russian Defence Minister told the Tass reporters that the USA was using Yugoslavia as a test range for its latest secret means of destruction. The existence of the weapon had not been denied by the USA. It has also come to our notice that some other countries like Taiwan carried out exercises to assess the capability of such a weapon in recent months.

The electromagnetic pulse is, in fact, an electromagnetic shock wave. The EMP energy produces such a storm of electromagnetic field that it produces short lived transient voltage of thousands of volts (Kilovolts) on exposed electrical conductors like wire or conductive tracks on printed circuits boards where exposed. The idea would be to immobilise the enemy before destroying him. Information flow is to be strengthened for own side and stopped for the enemy.

The weapon can be delivered by a cruise missile or by any other stand off platform and exploded within yards of a target. The explosion would be followed by generation of high power microwaves that can disable electronic circuitry in computers and communication equipment. This EMP could erase software and computers.

For a military man it wouldn't be an easy task to identify targets for the electromagnetic bombs because of the complexity of the entire system. It may be easy to identify certain targets but to match the bomb capability with the target may not be that easy. Buildings, housing various government departments, and installation areas are easy to locate and target because of their geographical locations and can be attacked provided the aircraft can penetrate to weapon release point and height. Mobile and camouflaged targets like air defence radars, missile complexes, mobile troops, naval vessels over the high seas are also good examples of the target system that can be attacked with E weapons. Like any conventional warheads, E bomb could be fitted on a cruise missile, conventional missile or on an aircraft.

However, like all other weapon systems, E weapons also suffer from their inherent limitations. Despite being an area weapon, the E weapons will demand better accuracy for their delivery because it would be important to illuminate the target area completely to achieve the best results.

Directed Energy Weapons:[11] Like EMP weapons, Directed Energy Weapons (DEWs) will revolutionise the military affairs. While talking about RMA, Kapil Kak of IDSA mentioned that,

" ...Technology change may well revolutionise warfare in the 21st century. Countries that can exploit emerging technologies and synergies the same with innovative operational doctrines and organizational adaptation could doubtless achieve far higher levels of relative military effectiveness.[12] "

That is why it has become necessary to discuss and understand the impact of DEWs. These weapons are at a very advanced stage of development and would be more dangerous and lethal than anything seen or experienced before. The directed energy weapons neither make any noise nor do they inform about their deployment. The super observation capability of a nation may discover the design of an adversary but it would be very difficult to determine the time and direction of the attack and the effectiveness.

A complete paradigm shift in military aspects would be expected when the Laser and Particle Beam weapons would be added to the military inventory. The cyberspace-based technologies supported by the DEWs in a future conflict might inflict unacceptable damage sufficient to alter the course of the conflict. The airborne laser system is being deployed by the USA recently with capability to destroy a missile with a range beyond 120 km at boost phase. The space based laser or particle beam weapon capable of punching a hole to any missile at a range of 3000 km is under active development. Nuclear bombs and the perceived threat vis a vis these new developments have to be analysed by the military thinkers and take effective action before being taken completely surprised. Nuclear doctrines, mutually assured destruction, limited response, CTBT, MCTR etc., will lose their relevance in the backdrop of the silent revolution of DEWs.

Precision Guided Munitions (PGMs): PGMs are among the most important systems to drive the RMA ahead. In India, some progress has been made towards infra-red or command guided missile systems, but Indian capability to produce or indigenise the systems like laser-guided bombs etc., is very limited. Only alternative would be to stockpile the same and ensure a steady supply at the time of crisis.

In the Gulf war it was estimated that ten percent of munitions were PGMs. In the war of Kosovo, the majority of US bombs were laser guided variety, which ran into trouble due to bad weather. On the contrary in Afghanistan the bombs were all weather and satellite guided. The JDAM or the Joint Direct Attack Munitions, a guidance kit fitted to the ordinary iron bombs, transformed it into an accurate munitions using satellite linked global positioning system to guide it to the target. Though it was less accurate than the laser guided bombs[13]. Yet, the advantages of the same are that these are much cheaper and can be dropped by a wider range of aircraft.

Command and Control: The thrust area for the Indian armed forces would be:

- Improvement in graphic display of terrain in terms of image capture, image processing, resolution and speed.
- Portability of imagery data from one medium to another with the desired degree of resolution.
- Data fusion techniques to cater for inputs from large number of different types of systems.

Electronic Warfare: Major effort also would be required in the following fields:

- Radiation detection and sorting in a dense signal.
- Stand off jammer for wide area penetration of enemy's electronic system.
- Extension of electromagnetic spectrum to millimetric and optical wavelengths.
- Warning of illumination by IR, Laser and Radar.
- Electronic protection of Command and Control systems.

Human Element of RMA: No revolution can ever be complete, without bringing the human element in the loop. Whether it is a man-machine match or sensor to shooter match. It all depends upon the individual to understand the complexities of the future scenario and act correctly and make no mistakes. That is why it is necessary to understand the main player behind all these developments. In addition to the capabilities and technologies inducted into war, attention must be paid to the development of the quality manpower for future military operations. Age old traditional military will be at a loss in this changed scenario unless it is ready to accept and learn the complexities of the situation. Human computer system integration is a vital requirement. That is why the future defence personnel will have to be well trained, and undergo a different type of training in addition to their traditional trainings, which may seem complicated at the outset. But the machinery, if fine tuned, would resolve the issue appropriately. They would have to control and work with machines and information system in the most efficient manner and be mentally and physically superior to the opposing forces. Considering the fast moving nature of the future battlefield, the human factor should not be a limiting factor for the commander; otherwise the RMA will be a lost case. As a result information overload for the human being has to be relieved by introducing voice recognition and voice generation, gesture recognition and responses. For a successful revolution in Indian military affairs, consistent human engineering will be an urgent requirement for cost effective human-machine fusion.

Conclusion

Despite living in an intense nuclear neighbourhood, India would be facing both low and high intensity conflict situations in this century. The presently acquired nuclear status of India wouldn't be of much help towards RMA as it stands today. No doubt, there have been some positive achievements in the field of Battlefield Damage Assessment and surveillance capability assisted by satellite imagery and induction of UAVs. A steady supply of PGMs has also been ensured as was displayed during Kargil conflict. DRDO has made some impressive gains in the field of defence technology. Yet it seems that the Indian military would have to work hard to integrate itself with the information driven digitised battlefield. There wouldn't be many choices left but to adapt to this ultra sensitive environment. High mobility and rapid acquisition would be the most potent parameters. The new weapon systems being developed indigenously are yet to be inducted and till such time these are made available with the military, the system cannot be integrated effectively in overall planning. Therefore it would be necessary to incorporate the new developments into the operational planning to bring about the Revolution in Indian Military Affairs, or else the Indian military would face a new type of battlefield previously unknown. While talking about the same, Thomas Welch in Institute for National Strategic Studies remarked as under:

"In USA, RMA was made possible by the information revolution, long-range precision weapons and simulation capabilities. We do not yet know the military organisations and operational concepts that will characterise this revolution.[14]"

It would be pertinent to suggest here that Indian armed forces could have their own laboratories manned by dedicated scientists to develop the futuristic weapons under a dedicated military/scientific leadership. It should be understood that technology driven RMAs are brought about by weapons or systems exploiting powerful combinations of technologies, combined with supporting doctrine and organisation. Input process being the technology breakthrough, and the output being the RMA. And that is why following may be the logical steps for Indian military to achieve Revolution in Military Affairs:

- Synergy with DRDO to develop the required technology.
- DRDO accountability for projects and cost overrun should be instituted.
- Enter into joint ventures with friendly countries for development of defence technologies.
- Take active interest in the development of EMP and Directed Energy Weapons.
- Ensure steady supply of PGMs at the time of crisis.
- Improve upon command and control structure of the three services, by ensuring steady flow of data, data fusion and interpretation.
- Develop human resources to match the technological develpments of the future.
- IITs and Indian Institute of Science along with DRDO should be funded to carry out R&D in the field of related critical technologies.
- MoUs with cost overrun /time overrun penalties should be instituted for every defence related projects.

Notes:

[1] Martin Van Creveld, *On Future War,* 1991 Brasseys, UK.

[2] Ibid pp2-3.

[3] Carl Von Clausewitz, *On War,* Princeton, NJ Princeton University Press, 1976.

[4] Edward Waltz, *Information Warfare Principles of Operations,* Boston/London 1998, Artech House Norwood MA. p133.

[5] Martin C Libcki, *What is Information Warfare?* Washington DC, National Defense University Press, 1995, p60.

[6] Air Cmde CN Ghosh, "Satellites and Space Application" *Indian Defence Review,* Vol 16 (2) Apr-Jun 2001.

[7] Ibid

[8] Air Cmde CN Ghosh, "Application of Unmanned Combat Aerial Vehicles in Future Battles of the Subcontinent", *Strategic Analysis,* Vol XXV No 4 July 2001, pp599-600.

[9] Annual Report, Ministry of Defence, Government of India 2000-2001, pp64-70

[10] Air Cmde CN Ghosh "EMP Weapons", *Strategic Analysis,* Vol XXIV No 7, October 2000, pp1333-1348.

[11] Air Cmde CN Ghosh, "Directed Energy Weapons" *Strategic Analysis,* Vol XXIV No 11, February 2001, pp2005-2070.

[12] Kapil Kak, "Revolution in Military Affairs", *Strategic Analysis,* IDSA April 2000.

[13] Ibid p2.

[14] Thomas J Welch, *Revolution in Military Affairs : Our Perspective* (Strength Through Cooperation:Military Forces in the Asia-Pacific Region) Institute for National Strategic Studies, pp9-10.

3

Military - Media Relations

John Street maintains that " media have power; they determine the fate of politicians and political causes; they influence governments and their electorates." And no doubt they can influence the course of present century's war by casting their influence on the military activities as a whole. The rapid advancement in technology has resulted in the media assuming a very significant role in accelerating and expanding collective awareness of events, issues and concerns. As was evident during the Kargil operations in India and Kosovo war in the Balkans, media played a very important role. The Chinese did not take Chinese embassy bombing by the US aircraft kindly and the Kargil reporting by the television media galvanised the entire nation. The 11th September World Trade Centre attack by the terrorists may be the latest classic symbol of media power. It could be seen that it ignites passions, sparks new perceptions, crystalises deeply held beliefs and compels people, nations, organizations and institutions to examine and define the contemporary issues. It is also instrumental in influencing public opinion and manipulating their perceptions.

The future military operations are likely to take place in an intense media glare, which will be both interactive and pervasive in its presence and influence. The CNN, BBC, Star and Aaj Tak correspondents were in Afghanistan covering details of every development and reporting through satellite communication system. At times this could compromise security. The top secret Somalia landing of the US troops under the glare of CNN TV cameras may be cited as an example. No doubt the operational details got leaked out but it was a lesson for the armed forces all over the world about the kind of future they will face in years to come. The media has thus become an awesome force or rather a force multiplier.

Our adversaries and their surrogate front organisations operating within the country will seek to gain an advantage by influencing the media think tanks, academic institutions, non-governmental organizations (NGOs) and international agencies. The websites maintained by the organizations like Jammu and Kashmir Liberation Front (JKLF) and other terrorist organizations are the live examples that how this new type of media blitz can put pressure on the entire governmental organisations even in peace time. Agra summit is also another classic case, where a proactive media policy could have avoided many embarrassments for the Government and the nation as a whole. Media driven perceptions and opinions will have significant bearing on the decision making process at the strategic level and nature and scope of military operations at operational level. Modern technology enables a greater number of individuals, groups, organizations and nation states to be linked to the world through global information networks and they are likely to pursue their interest by attempting to manipulate and influence the media.

Apart from being an eye to see everything, the media plays an important role in psychological operations, image projections in the prevailing environment of low intensity conflict that India is fighting since its independence. The media, particularly regional and vernacular press, is being expertly and extensively exploited by the anti-national elements as well as their front organizations to discredit, defame and pressure the armed forces to slow down the tempo and intensity of the operations. Therefore it is necessary to take a holistic view on media-military relationship of the future and re-organise the existing organisation. Policies and procedures require to be revamped to ensure effective dealing with such an important element like media. Media and the military must understand each other's requirement and try to fulfill it towards a common goal to enhance national security.

There are serious shortcomings in the existing media policy of the Indian military. And if it were accepted that media could be a force multiplier then it would be necessary to examine the existing organisation, policies and procedures and suggest changes.

Existing Organisation

The Press Information Bureau of the Ministry of Information and Broadcasting is the authorised agency for providing information on the activities of the government to the representatives of the media. The Directorate of Public Relations of the Ministry of Defence functions as a Defence wing of the Press Information Bureau. It acts as a link between the Ministry of Defence and the citizens to provide information to the mass media. The Director is an officer of the Indian Information Service and his rank was recently upgraded to the equivalent of a Joint Secretary in the IAS or Maj Gen in the Army. Despite this change DPR comes under the control of JS (G) in the Ministry of Defence and his authority has been curbed very seriously. There are about 24 PRO units under the Directorate of Public Relations located at various media centers all over the country. The primary function of this organization is to take care of public relations on ceremonial and other occasions and arrange press visits, organise press conferences and issue press releases and rebuttals, if at all. This Directorate has been organised basically to provide routine kind of media coverage to the ceremonial and peacetime activities of the Services.

Such an organisation cannot rise to the occasion as was found out on different occasions. Its bureaucratic approach and hierarchical pattern prevent it from reacting quickly to developing situations. Moreover the head of this organisation is an officer from the Information and Broadcasting Ministry who does not have the right type of background to become an interface between the military and the media. During Kargil operations the ineptness of the organisation was very clearly pronounced. As a result tri-Service daily briefing presided over by JS (XP) of the Ministry of External Affairs became the only authentic source of information for the media. This organisation was left to carry out only the administrative arrangements. The organisational head is marginally familiar with the subject for which he is responsible. He himself has very little authority to interact with the media on other than routine matters. Any matter of importance requires clearance at higher levels, through step-by-step approach. The existing media organisation is hence least responsive to the proxy war scenario of the present day. As a matter of fact Indian military has been fighting low intensity conflict in many fronts but the media-military relations not being set at the correct level, generally enemy gained ground and Indian military was always at the receiving end despite doing a heroic job wherever they were deployed. This organisation is yet to understand the Net war being carried out by the many Pakistani terrorist groups and maligning the Indian military and the Government everyday. But unfortunately no response team has been put into place to wage this new kind of warfare.

Existing Public Relations units are inadequate. These are short of manpower and manned by low ranking de-motivated officers considered not good for their mainstream. Best example may be the number and inadequacy of these PROs in the Kashmir sector. The whole world is only fed with the information about the killing in Kashmir, which is not only demoralising for the troops but as well for the entire country. Loss of face in the international community is the obvious result. PROs, as stated earlier, are mostly unsuitable for the job, lacking the required training and experience to deal with media and at times lack the knowledge of their own service, its plans or programmes. Influencing public opinion and image projection are also equally important in the present information driven environment. Present arrangements are inadequate. These functions cannot be performed effectively by the individuals not trained adequately and do not have sufficient knowledge about the military matters

The requirement of the news hungry media is generally not understood by the military. A reporter dispatched by any renowned media organisation to a conflict zone would like to get stories on a regular basis, which can be transmitted to the whole world. But the commanders or their representatives, who are engaged to interact with the media, do not understand such urgency of correspondents. It is necessary to feed the media constantly because they cannot afford to waste their time and money on an individual spending time with the troops and not reporting anything worthwhile. Unless there is a well-orchestrated plan before hand, it would not be possible for the commanders to react to a situation under enemy attack. The media-military interface has to be a part of every operational plan and exercised regularly like all other military plans. It should be understood very clearly that at no time and in the heat of the battle, operational plans couldn't be disclosed or indicated. Greatest care must be taken to deal with the electronic media. Any mistake or lack of discretion will compromise the security of the location of the unit, which will largely jeopardize the overall attack or defensive plans. During Kargil conflict certain media personnel were allowed to televise the position of certain bunkers etc., which was not a good example of media publicity because the bunker location was found out by the enemy and came under heavy enemy shelling. There is a general lack of experienced media personnel on defence matters or of war correspondents. Generalists pressed into the service in a conflict situation cannot be expected to report correctly under such pressure.

PROs located at important cities do not function under the respective Hqs/Corps/Commands and in fact report to the DPR in Delhi. This often results in lack of coordination between the PRO and the higher formation headquarters. Similarly, the Service specific PROs located with the Directorate of DPR do not have authority over the PROs located at the respective Command Hqs/stations, who report to the DPR directly. Current and emerging electronic technologies permit any aspect of military operation to be made known to a global audience in real time and without the benefit of filters. In the low intensity conflict scenario the information has to be provided to the media in real time basis by the formation deployed in the operational areas. The existing network of the PROs does not cater for the operational requirements of media at the divisional or brigade levels. The PROs presently do not perform which they are required to represent and project and hence the dichotomous situation.

Handicaps of PROs: Basic media related infrastructure is not available with the PROs such as STD authorized telephones, FAX, high speed internet connection, suitable transport and accommodation. Moreover, public relations has become a command function because in the low intensity conflict and counter-insurgency environment, the field commanders have to interface personally with the local population with a view to win over their hearts and minds and influence their opinion in favour of government policies and operations and against the militancy and the militants. The public relations thus becomes an important command function and an operational imperative which cannot be performed through the existing DPR organisation. The commanders at all levels need to have necessary training and organisation to discharge the important command function of positive and personal interface with the local population.

Defence Technical Publicity Rules: The Defence Technical Publicity Rules were formulated in 1982 primarily for the new entrants to Directorate of Public Relations. These rules were formulated when Army was not so actively involved in the counter-insurgency operations and the concepts of low intensity conflict and proxy war were relatively unknown. The invasion of the satellite channels was also not visualised at that time. There is an urgent requirement of revising and updating these rules with a view to make them an effective instrument of proactive use of the media as a force multiplier for psychological operations, operational publicity and image projection of the services.

Healthy Military-Media Relations

Media management is a very commonly used term in the armed forces. But it should be understood that media cannot be managed. It has to retain its independence and effectiveness and military has to understand its impact. A military commander will have to think twice before firing a shot and that too in a war like situation. The present system is inadequate to fulfill this task and the system of PIB controlling public relations of the armed forces should be revised. Therefore, it would be perfectly in place to have DPR attached to Ministry of Defence (MOD) to cover all the activities related to it only. The Service should have the autonomy to manage their own public relations, psychological operations and image building. This necessitates positioning present PROs in their respective headquarters.

It would be better for the Services to create a Joint Services Media cell which should function under the Chief of Defence Staff (CDS). Coordination of the inter-services activity would be the main job of this cell. On the other hand, each of the Services Hqs should have their own Media cells headed by an officer of the rank of Colonel or equivalent. A professional soldier trained in media matters should be heading this post, who would coordinate all media related matters pertaining to individual Service. Similarly Media-Military Relations cells should be established with all commanders at the Command and Corps level for the Army and for Air Force, and for Navy at the Command level only. This cell can be positioned at the active bases under the Base commanders in case forward bases like Srinagar and Avantipur are involved in Kargil like operation.

Advanced communication facilities are required to be placed with these cells. During Kargil operations, all TV films were sent by road to the nearest airfield and airlifted to Delhi for telecast. Present generation communication helps to uplink the TV shots directly from the place of activity for telecast. Such equipment should be available with the forward formations, which will greatly enhance the reporting capability of the reporters.

The officers handling the publicity must have the authority to interact with media, issue rejoinders/rebuttals and organise press conferences, etc.

Technological advances during the last decade and the increasing globalisation of news reporting have created a unique situation where news reports can be transmitted live to an international audience. The present terrorist attack in the USA may be taken as a classic example. Every minute detail is known or seen by the TV cameras. And that is why the military must understand that control of media product is not feasible. Gulf war has become a benchmark for many a thing for the military and media participation. The theater commander had been giving the privileged briefs to pool journalists on future operations. Indian military can take a leaf out of these experiences to suite the Indian context. However, it cannot be denied that military-media relationship should be based on more frank and open relationship in which military should accept a degree of healthy and justifiable criticism. And journalists on their part should restrain themselves from negative and disruptive reporting. Dramatising an insignificant event may project a journalist as hero to its editor. On the other hand remaining too circumspect on everything else will project the military in a poor state. Both the parties should understand that apart from their individual growth and pride, national psyche is deeply involved in a situation like war.

Another challenge for Indian military would be to understand the media criticism. Any publicity with an undertone of criticality is either not understood or taken very kindly by the military brass. They consider such an act as breach of trust. And it is often alleged that media is obsessed with negative reporting to enhance their own image and call it an objective reporting. Outright denial of such an activity in Indian environment may not be possible but military should not over react and criticism must be handled with an open mind and a sense of equanimity. On the part of the media they should also understand the inner functioning of the military establishments. Any story filed on only hearsay evidence and not well researched could not only damage the military repute but also can cause severe damage to the personnel in charge of these departments in the eyes of the nation and other compatriots. And if the officer is in the promotion zone, it may hold back his carrier progression. It should be understood, that a reporter has the most powerful press behind him who could flash his story internationally but a military man does not have any resources for rebuttals. He can only suffer silently and suffer the consequences both officially and socially. A Court Of Enquiry is generally ordered as a follow up of any malicious report. Even if the officer is proved not guilty, the press will forget to carry it. Here lies the greatest distrust between the press and the military. The media must understand the cause and effect factor of military reporting. But military must understand the reality. As Gen Arjun Ray pointed out," Indian media is becoming more competitive and consumerist. News is like consumer commodity and people are hungry for sensationalism. Media caters to the taste of the readership; security forces must appreciate that. They need to come to terms with this fact of life[2]."

Media support to the Indian military forces during Kargil operation helped the military gain tremendous support from both national and international audience. India's stand on LOC was clearly understood by the international community. Media so harnessed could be beneficial to a military campaign provided the field commander could play his role correctly based on a healthy " I understand and you understand basis." It may be understood that a correct relationship between military and media supported by the objective rules and regulations, could be gainful for both the parties.

Need for Media Specialists

Much of the problem emanates from the media personnel also. In general it has been found that in India there is a general lack of interest among the reporters to specialise on defence reporting. They would prefer to report on politics or economics rather than working on defence matters. Probably except during war or war like situation, defence always takes a back seat and there will be a lack of readership even if they are reported upon. As a result most of the defence reporting during peacetime relates to ceremonies and grandeurs. Therefore, whenever specialised defence related event requires to be reported upon, media banks upon the retired defence personnel. That is why "journalists have to be trained and provided with resources, they depend on access to technology and communications infrastructure[3]."

Regional media, mostly the vernacular press, face a peculiar challenge. Barring few, there is a general lack of understanding of the military matters by the regional correspondents. As a result, military reporting suffers. Situation even could be worse when such correspondents report on military operations carried out from their part of the area. Vernacular press could create many difficulties for such military operations[4].

As a spillover of the colonial mentality and the much feared "Official Secrets Act" Indian military has completely estranged itself from this powerful force multiplier of this generation. It should be understood that it is not possible to keep the secrets like the yesteryears. Military has to be more transparent and prompt in informing the media. Holding back information will become a self-defeating exercise. Military commanders refer to Special Army Order (SAO) 15/S/81 and the Army Rule (AR) 21 to prohibit contact with the media. But if read carefully it may be seen that the orders are not that stringent as it is made out. Yet the order persists, that defence personnel are not to communicate to the press any matters connected with India's defence matters without MOD clearance[5]. A complete course correction by the military will be right in tune with the requirement of this century.

Conclusion

"The Inter-American Press Association has criticised US media coverage of September 11 and its aftermath, warning that its credibility is at stake, following a report surfacing about the five US television networks CNN, NBC, ABC, CBS and Fox who agreed to national security adviser Condolezza Rice's request to exercise caution while airing statements by Osama bin Laden and Al-Qaeda[6]."

CNN despite its wide coverage is not showing anything anti-American or examining American policies during this present terrorism and Afghan crisis. Indian populace on the contrary have access to many more sources of information, instead of being served by the national media alone[7].

"The familiar claim that knowledge is power captures the most commonly cited connection between mass media and power. The idea that knowledge is a source of power and that mass media are central to this rests on some fairly straightforward if not uncontroversial claims[8]". The military-media relationship has to be based on such central ideas. On the part of the military, it should be understood that, " Devolution of information dissemination system should be the key to the military's interaction with the media. At present there is little interaction between the staff and the media at the command, corps level. Any press release which has to be issued is required to be cleared by successively higher authorities[9]." Media cannot take a back seat any more in matters pertaining to military. Military commanders must educate themselves to deal with media effectively and use them to their advantages. At the battlefront, they assess the requirements of journalists and feed them as much information as possible so that they can file their stories on time keeping the secrecy of the location and military plans. Judicious use of the information will go a long way to establish a complete rapport with the media. An official spokesman at the level of Command, Corps, Division and Brigade level should be a step in the right direction. In addition, periodic briefing by the heads of the institutions would go a long way to keep the media updated on the country's security matters.

On the part of the media they should understand the need to build up expertise on defence reporting. There is a requirement to train different levels of media persons both as defence correspondents and analysts to maintain its credibility. Readership or viewership will reduce drastically if the reporting is not objective. Viewers will turn to foreign source for information rather than depend on unreliable national media. And once the credibility is lost, it would be an uphill task to restore the viewership.

Notes:

1 John Street, *Mass Media, Politics and Democracy*, Palgrave, New York, 2001. p231.

2 Maj Gen Arjun Rai, *Kashmir Diary*, New Delhi, Manas Publications, 1997, pp83-85.

3 Ajay K Rai, Military-Media Interface: Changing Paradigms, New Challenges, *Strategic Analysis* Vol XXIV No.5 August 2000, p902.

4 Ibid p915.

5 Ibid. p913.

6 *The Times of India* October 19, 2001, p1.

7 Ibid

8 n1, p233.

9 n3.

4

Roles And Missions Of Indian Armed Forces

Certain academics club the roles, missions and functions of the armed forces and explain away by stating that though the words are different but overall meaning remains the same. These words only describe what the services do. But such limited reasoning do not bring out the correct and specific meaning. Because, "**Roles**" are the basic purpose for which a service has been created, organised and trained for sustained defensive and offensive operations. Whereas, "**missions**" may be the specific functions of the commander of a particular force. While examining the basic roles and missions of the armed forces, it comes out clearly about the battle space overlap and its ramifications. It is the result of technological advancements; advancement in electronic warfare techniques, application of the 4th dimension to the future warfare contributed by the satellite and space technology and warfare in cyber space. No doubt, there will be general dilution towards understanding of "Roles and Missions" of the individual services of this generation. The changes in geopolitical situation, declining budget and "threats to institutional relevance[1] " will also create further confusion to understand the definitions correctly. This may be a pointer that while describing the "**functions**," President Truman stated in 1947 that,

" Functions of the Armed Forces, ... include those various activities, operations, and capabilities for which the services were responsible and for which they were charged with the organizing, training and equipping of the forces[2] ."

But at a later date the functions were described in greater details, which distinguished between primary and secondary functions, by a US Department of Defense Directive (DoD) 5100.1 which specified 17 primary functions, four collateral functions, responsibilities related to space and responsibilities concerning combat operations in support of other services[3] .

Such views need to be analysed in Indian context to understand roles and missions of the Indian armed forces. Indian armed forces are not immune from the global, technological and social changes facing the entire mankind. Terrorism is raising its ugly head in different parts of the country. Neo-globalisation compounded by technological advancements and information revolution would have their collective effects. Society will not be ready to absorb many impacts brought in by these ultra new factors. Military power and the military balance in the third world countries like India will be at crossroads, unless such changes are understood and absorbed quickly to define future roles and missions. Factors like growing interdependence, spread of market mentality, narco terrorism, insurgency, low budgetary support, security dilemma and lack of understanding about the security needs and functioning of the Services have to be taken into consideration towards such analysis.

India has the tradition of democratic governance since independence and it has fully met its international obligations whenever called for. It is expected that such democratic foundation will remain steady in future despite internal schism on many fronts. Armed forces in India remained totally apolitical and are expected to do so in future as well. It may be stated that India presents a successful model of democratic development in the entire developing world. Despite fighting four wars and numerous attempts of destabilization of the nation over the last decade and a half, the country remained unshaken. Yet the threat from nuclear China and Pakistan are real. There are other destabilising factors, both internal and external, which would thrust upon many responsibilities on the Indian armed forces to maintain stability of the nation. Capability and flexibility must be there to meet the future threat to the territorial integrity and sovereignty of the country. And "Roles and missions" are the direct fallout.

Threat Assessment

India like any other nations cannot hope to manage its security without actively participating in international and regional security issues, which emanate from global factors, trends and developments, compounded by domestic uncertain situations on many fronts. Analysis of international, regional, and domestic scenarios would only correctly orientate ourselves to our security concerns and that would lead to development strategies to safeguard Indian national interest, predominantly guarded by the Indian armed forces.

Indian Geography: India has the unique geographical position with 3244 km land border with Pakistan, a 4056 km difficult and disputed border with China and a 4351 km border with Bangladesh. While India has borders with seven neighbours, a few of them share borders among themselves, which brings to focus our special relationship with all these neighbours. Moreover asymmetry of size, population and economy in relation to these not too friendly neighbours also creates many prickly problems which do not have any ready made solution. Inferences of Indian regional hegemony is resented by most of these neighbours, and Pakistan clearly defies the theory totally and intends to have a level playing field with India among the comity of nations.

International Environment: Viewed from a larger canvas, India is seen as part of the Southern Asian region which stretches from West Asia through Central and South Asia, to South East Asia. This entire region is a part of an area with a high intensity of trading through Indian Ocean. Strategic backdrop of this entire region includes nuclear weapon states or nuclear capable states, starting from North Korea, China, India, Pakistan, Kazakhstan and Israel. All these countries are with missile capability with radius of action covering the entire length breadth of this country. Over the years this region has been witnessing an unprecedented proliferation of terrorism and narco trafficking which in turn threatens the stability of states and societies.

On-going US armed intervention in Afghanistan to contain terrorism may not augur well for the entire region. Being an unwilling participant in the Afghanistan crisis, diligently fostering Jehad culture for many years, and the presence of Pushtooon refugees on its soil, will create stability problem for Pakistan. A high level of uncertainty and unpredictability marks contemporary strategic and security environment. The international order is evolving into polycentricity with a number of centers of power, each with different level of strength in the attributes of power. This period of flux and uncertainty is likely to continue deep into this century.

While the risk of organised war has perceptibly reduced in the overall sense, the threat of irregular, especially proxy war linked to social conflict, has increased. Most countries are likely to experience continuing domestic socio-political turbulence. Ethnicity and religious resurgence in many cases have given rise to a new sense of nationalism, which generates centrifugal pressures. Correspondingly, there is a dramatic rise in the number of non-state actors, such as crime and drug cartels, religious fundamentalist groups, the free availability of technology, Internet revolution, radical ideas and militant cranks, etc. Transnational terrorism with an active support of some nations is turning out to be the main scourge of the 21st century. The 11th September WTC catastrophe, which took the lonely super power by surprise, is indicative of the type of warfare that the nations will be facing in this century. The external linkages of state support aggravate the problems of societal cohesion and state security.

The Indian Ocean serves as a maritime link between the Atlantic Ocean and the Mediterranean Sea on the west and Pacific on the east. Major sea-lanes of the world pass through this ocean. Therefore, the security of Indian Ocean is also important for the economic growth of Asia-Pacific region in general and India in particular.

Regional Environment: The economic reforms and resultant upsurge in growth has raised the economic capabilities of China manifold. China is a declared nuclear power with an impressive inventory in all manifestations of that status. People's Liberation Army (PLA) is the largest army in the world. China's aspirations to fill the vacuum created by the former Soviet Union wouldn't be entirely misplaced.

The strife in Afghanistan will have a direct impact on the Indian subcontinent with the impending resolution. Narcotics and illicit arms trade will get a boost in the region, which in turn will have direct implications on our national security. An unstable nuclear Pakistan on the other hand will add to the Indian security risk many times over.

State sponsored proxy war with finances through narcotics and drugs, state protection and even the use of military forces to augment the turmoil in the border areas of the country would add to the already volatile security situation.

Internal Environment: India faces several internal challenges as it attempts to develop and seeks to gain a rightful place in regional affairs. Poverty and population are primarily responsible for slowing down the country's modest growth added with political instability and absence of a strong central government. Illiteracy also remains another major hurdle for social upliftment and checking population growth. Cross-border terrorism, external interference, ongoing insurgencies in the North Eastern part of the country, and Sri Lankan ethnic clashes spearheaded by Tamils of that island state will pose many security related problems to this country. India is an energy deficient country and over dependence of oil import will have many security ramifications

Specific Threat from Pakistan: Before the World Trade Centre destruction by the terrorists, the world community, and in particular the USA, refused to take notice of Pakistani active involvement of encouraging terrorism against India. Taking advantage of an open democracy and a pluralistic society, Pakistan could over a period of time spread terrorist activity throughout India, including North-East, where separatist movements are being instigated. Destabilising India through actively sponsored terrorism, had been the state policy for Pakistan for a long time and is likely to continue well into this century. Pakistan has emerged as an epicenter of Islamic fundamentalism, inducting Islamic Mujahideens into Afghanistan in the north and Kashmir in the east. Pakistan had been the creator of Taliban regime of Afghanistan, and together they spread the terrorist activity directed initially mainly towards India and followed up by attacks on American lives and property round the world. The final act of the terrorists was seen on 11th September, 2001, when the world was shocked by the most heinous act of terrorism ever seen or contemplated in the history of mankind. The USA and the world grudgingly accepted the role of Pakistan in the terrorist activity and finally the USA brought its force to bear on Afghanistan. Pakistan became an unwilling ally with ferocious protests from the fundamentalist groups around the country. At present, things in Pakistan do not augur well for that country, which in turn would cast their shadow over Indian security matters. Pakistan has done enough to draw India into Low Intensity Conflict over a period of a decade and a half in Kashmir. But it may be pointed out here that it is not a problem of Kashmiri Muslims alone. This will encompass the other 120 million Muslims in India as well. However, Kashmir is not the only cause for Indo-Pak tensions, but is a manifestation of other basic insecurities in that country. The US punishment of Afghanistan, and the US determination to keep its permanent presence in Pakistan despite local hostility will have serious regional impact. Dragging along an unwilling Pakistan to fight terrorism which it had garnered for many years, will cause many unpredictable changes in the geopolitical balances in the subcontinent which cannot be predicted as of today. However, embroiled as it is with internal turmoil and economic instability, burdened by million Afghan Pushtoons, Pakistan would always have a strong anti-India bias in future. Complete rout of the

Taliban by incessant American bombing and establishment of the Northern Alliance control over Afghanistan, and President Musharraf's complete turn around at least on paper won't change the Pakistani mindset. The 12 January, 2002, and the 27 May, 2002, addresses to the nation by the Pakistani President repeated the typical mindset of the entire nation. He almost repeated his speech of the earlier date continuing to support Kashmiri cause and prepared the nation to listen to him by carrying out the test firing of Ghauri and Ghaznavi missiles in May 2002.

Therefore, it should not be forgotten by the Indian strategic community that Pakistan is a nuclear power and is very unstable politically. As such brandishing of Pakistani nuclear arsenal should be taken seriously and all military planning should include this real threat of the present and future.

Threat from China: China occupies more than 40,000 sq. km of Indian territory, excluding a large area in the north illegally ceded to it by Pakistan, and lays claim to another 90,000 sq.km. Chinese desire to fill the vacuum left by the former Soviet Union is well known and India is no position to contest such claim. India has to understand that there is a clear asymmetry of conventional as well as nuclear forces in favour of China. China's growing economic and technological strength is also creating asymmetry in certain other dimensions. That is why, it should be clearly understood that China poses both direct and indirect threat to India. The direct threat is related to resolution of the Sino-Indian boundary dispute, whereas, the indirect threat could be through surrogacy of our smaller neighbours and growing economic strength in International market forces. In recent years there has been a certain shift in Chinese posture towards India after the nuclear tests.

China has embarked on an ambitious modernization programme and is acquiring state-of-art weapon systems for its army, modern combat aircraft and in-flight refueling capability for its air force and improving its large submarine fleet. China has also a very ambitious space programme and would like to counter the US space militarisation as far as possible. Chinese Directed Energy Weapon development programme should also be taken note of. In Tibet, China is improving existing infrastructure, which will enhance its potential to deploy large military forces on our northern border.

China has had a close defence relationship with Pakistan for many years. China never bothered about international protests against its transfer of nuclear and missile technology to Pakistan. M11 missiles supplied by China can only be targeted to India. Chinese overt and covert acts of supplying hi-tech defence equipment and arsenals to Pakistan have the only obvious strategic overtone, of using Pakistan to keep India engaged militarily. When we take these factors into our strategic calculations it becomes clear that we cannot afford to ignore the seriousness of a situation that could result from a sudden deterioration in relations with China. China, no doubt, is a crucial factor in India's security calculations.

Maritime Threat: India is a maritime nation. Apart from our long coastline and far flung island territories, which have security implications, India's overseas trade is considerable and as such freedom of the seas is essential. Our entire oil import comes from the Gulf and if this is hindered the country's survival will be critically affected. Safeguarding of resources in our EEZ, maritime assets and infrastructure is very crucial for our security dimensions. A credible ocean security apparatus by Indian Navy will be the requirement of this century.

Indian Security

Definition of security has narrowly been understood as protection of territory from military attacks by outside forces. The concept is actually open to broader interpretation on the face of variety of unconventional threats like environmental degradation, transnational organised crime, uncontrolled migration, and potential damage due to international market forces, narcotics, terrorism and militancy. India's security has to be viewed against such a backdrop, which is nothing but an integral component of its development process. India continuously aims at promoting an ever-expanding area of peace and stability around it, with a desire for the development without disruption.

With nuclearisation of the subcontinent, and China being recognized as a nuclear weapon state with credible delivery capability based on IRBMs and ICBMs, it is expected that conventional warfare thresholds are less likely to be crossed. As such, in this century India must prepare to fight a well orchestrated hi-tech battle against insurgents who possess modern weapons, explosives and means to deliver them, and operating in a networked field with state-of-the art communications. The armed forces would have to reorganise to face such situations. In particular, army must retain the primary capability of fighting a conventional war with a nuclear backdrop and reorganise itself along with the central police organisations and paramilitary forces into a cohesive and cost effective Internal security mechanism with a clear and unambiguous command and control, and logistics system.

National security is the protection and sustenance of core values and core interests crucial to India from external and internal threats, challenges, risks and vulnerabilities. Core values are clearly enshrined in our constitution. The national interests have to be protected and enhanced to ensure national security. National security thus constitutes institutional stability, social cohesion, economic development and military power including strategic nuclear deterrence. Defence policy is only one facet of this larger national policy and can only succeed with a harmonised balance within the grand national strategy. Therefore the national security objectives would be:

- Indian security forces must deter any external aggression who could try to gain control over Indian interest in any sphere, whether by land, air or to our Island territories.
- Indian scenario is hemmed by multi-ethnicity and multi-lingualism which has created many internal conflicts from the day of our independence and it would be necessary to secure an internal environment whereby our nation state is insured against any threats to its unity from religious extremism, or socio-economic dissonance.

- India as a member of the world community may particularly feel proud about its commitment towards a safer global environment by contributing effectively towards UN peace keeping and humanitarian causes. And it would be to the best of Indian interest to do so in future whenever called upon by the United Nations.

Roles and Missions of Armed Forces

Roles and missions of the armed forces would naturally flow out of the national security perspective discussed above. It is imperative that the Indian armed forces are adequately structured and equipped to meet the challenges of the present and future. All conceivable contingencies related to national security matters would be the direct responsibility of these highly organised and trained forces. Such imperatives dictate the **roles** of the defence forces:

- Defend the country, shore installations, economic assets and maritime trade against all possible threats including threats from nuclear, biological and chemical (NBC) weapons.
- Defence forces must be strong enough to fight any aggression from Pakistan or China or a possible collusion between Pakistan and China.
- The smaller neighbours of India somehow always identify India as a big bully. Any of their internal problems become a platform to point a finger towards the regional big brother to gain political mileage. Gujral doctrine could be of little help in many an occasion. That is why suitable contingency plans must be ready to extend assistance to Nepal, Myanmar, Sri Lanka, Maldives, Mauritius and Seychelles.
- For India's security interest, it would be necessary to save Bhutan from a Chinese attack.

- It would also be necessary to maintain special forces for the execution of quick unconventional Maldives type operations.
- Both the Indian neighbours (Pakistan and China) are nuclear powers and unless India can absorb a first strike and respond with an assured retaliatory conventional/ nuclear strike against both China and Pakistan, they will have unnecessary advantage over India.

While talking about traditional and non-traditional roles Samuel Huntington stated that,

"The distinction to be made is not between traditional and non-traditional roles but between military and non-military roles or between combat or non-combat missions. The purpose of the armed forces is combat: to deter and defeat enemy....That is their principal role or raison d'etre....Forces created to perform that role however, can be- and have been throughout our history – employed in non-combat, non-military role[4]."

In view of the above, it may be said that while the traditional roles of the armed forces will continue, the Services will have to continue perform the non-traditional roles like:

- Indian military had been in the midst of low intensity conflict (LIC) since Indian independence. Starting with Naga insurgency, Indian military is now facing an uphill task to contain cross border terrorism in Kashmir valley. Despite accepting heavy losses on manpower and equipment the armed forces are supporting civil authority to counter LIC. Unless and until a definite political will is displayed, the Indian army will continue to fight this low intensity conflict for a long time to come.
- India has always been giving support to the causes of United Nations and as indicated earlier it would be in the best of interest for India to continue to do so.
- As Huntington stated, civil authorities will always bank upon armed forces to help them during national emergencies like cyclone, flood or earthquake, because of their organizational strength.

These factors will relate to the fact that the military posture that demands maintenance of a force having both the requisite strength and capability to provide the deterrence. It will automatically call for enhancement of military capability on the face of such evolving threats and challenges. Unless this is done, the country's ability to achieve the national goal will not be possible. On their part, armed forces would have to focus on the following factors:

- Exploit Revolution in Military Affairs (RMA) in order to improve upon the ability to perform the future tasks effectively.
- Modernise older systems and incorporate present generation technology with an eye to incorporate the futuristic developments.
- Self-reliance and indigenisation for the military equipment is extremely important because high level of military technology would be very costly to import and sustain in future.
- India is no doubt a industrial power. Its industrial base has spread in different spectrum and it would be advisable for the military to take a note of the same and reduce foreign dependence as much as possible.

Missions of Indian Army

Indian army has the primary responsibility to guard a land frontier of over 11,000 kms. This includes a live and hostile border with Pakistan, which extends over 3200 kms and the difficult border with China, which extends over 4351 kms. In addition, the army of India has been constantly been engaged in internal peace keeping and counter-insurgency operations since independence. As a matter of fact Indian army may be the most experienced in the whole world on both these counts. Indian army has been contributing to the UN peace keeping most willingly whenever called upon to do so. To meet such diverse requirements army will have to develop the following:

- The present surveillance capabilities of the Indian army are not sufficiently strong enough to guard thousands of kilometers of border, which runs mostly along difficult terrain. Unless effective surveillance capabilities are created at the earliest, Kargil episode could reoccur to the much embarrassment of the national leadership.
- Army has to assess whether it has the capability to fight non-linear battle in future, in order to engage the enemy both at intermediate and depth echelons simultaneously. If not, then it has to build up the capability to fight the wars of this century effectively.
- Needless to state that, to achieve quick responses, army should acquire greater mobility and greater firepower in both offensive and defensive formation.
- Enhance mobility in the plains and in the desert so as to exploit opportunities effectively.
- Good amphibious capability to counter any threat to our Island territories.

Consistent with the threat assessment, future mission of the Army could be divided into following three segments:

China: Chinese economic and military superiority over India has been clearly established. That focuses on the Indian future responses to every conceivable Chinese design as follows:

- While considering Chinese threat from current force level in Tibet, the factors like present and future development plans of Tibet should not be missed. Chinese military would pose much larger threat to Indian territory once such infrastructures have been made available.
- Army would have to continue to develop infrastructure and logistics capability to support deployment of larger force levels and ensure high mobility all along the Himalayan border, to counter the Chinese thrust following infrastructural development in Tibet area.

- Heavy border deployment all along the Northern borders may have to be continued to ensure effective border management and prevent any further Chinese encroachments of Indian territory.

Pakistan: India has a total of 3200 km border with Pakistan and has the unique geographical characteristics of facing such hostile force from the high snow-capped mountain peaks of Siachen glacier to the deserts of Rajasthan. Pakistan would continue to oppose Indian hegemony in the subcontinent at any cost. There are sufficiently good reasons for Indian army to deploy specially trained forces and specialist equipment to face a hostile Pakistan all along the highly sensitive border. Three sectors have been identified for easier delineation of missions of the Indian army:

Jammu and Kashmir: Pakistan considers that its existence depends upon keeping the Jammu and Kashmir situation alive. It has been resorting to cross-border terrorism during the last decade and a half to keep the Indian army engaged in an unequal war. Indian army does not have any other choice than to adopt a posture of overall deterrent. The objective would be to:

- Deny Pakistan any territorial gains across the Line of Control.
- Prevent any future Kargil like situation.
- And plan offensive to capture maximum territory in Pakistan Occupied Kashmir (POK).

Plains of Punjab: Pakistan would not find it very fruitful to carry out any aggression into the plains of Punjab. It is expected that the plains of Punjab will remain relatively calm because the borders of the both the nations are fortified very heavily. Swift armour movement also wouldn't be possible because of the terrain and such heavy defences. The idea would be to defend territorial integrity in this sector. And as no war can be won by adopting only defensive posture, limited offensive should be planned to capture certain enclaves or gain tactical objectives.

Rajasthan Desert: Rajasthan desert could be the future battlefield of the subcontinent considering the terrain, defences and mobility of both the armies. Therefore, Indian army would have to:

- Defend territorial integrity and prevent Pakistan to create any launch pads to launch an armoured offensive.
- If Indian army has to prove its mettle and hand over a deciding defeat to Pakistan, then it would be the sector to plan offensive, to secure maximum territory and or destruction of Pakistan's strategic forces in order to force it to abandon her design on Kashmir and to negotiate from position of strength.

Island Territories: India fortunately did not have to fight any battle to safeguard the Island territories so far. But that does not mean that the preparation to defend these territories have to be neglected. That leads towards creation of amphibious capability of the Indian army. It has to create a typical force structure to dissuade adversaries from any adventures in our Island territories and create ability for quick deployment of that force in conjunction with Indian Navy.

Internal Security: Since our independence Indian army has been deployed to carry out the non-traditional military duties and these have never been cited to justify maintaining such a large standing army. The overall size, composition and training and equipment of the services have been based on our national threat perspective and combat missions to be performed. If the experiences of post independence may be of any help to delineate the roles and missions of Indian army then it should be prepared to:

- Combat long drawn cross border insurgencies and proxy wars in Kashmir and in the North Eastern part of the country.
- Assume responsibilities from paramilitary forces, or state police forces for the maintenance of internal security.

- Aid to civil authorities at the time of national disasters like earthquakes, cyclones and floods.

Indian Navy

The Indian Navy has to primarily defend the long coastline and the Island territories of Andaman & Nicobar and Lakshadeep as also to keep the trade route leading to our ports open. Maritime forces, because of their unique operating environment, can be effectively used to further national interests be they economic, diplomatic or strategic. The navy is a far ranging and flexible instrument of a nation state which can be used in an incremental manner to impose, coerce, or demonstrate support, wherever and whenever necessary. That is why naval missions can be combat, constabulary or diplomatic in nature. Therefore missions of the Indian Navy will be as follows:

- **Defence of Coastline and Island Territories:** Indian navy has to provide security to the coastline of mainland India and the Island territories and offshore maritime assets.
- **Protection of Maritime Trade:** Safeguard the maritime trade and especially make sure about the continuous flow of oil from the Middle East.
- **Operate Against Hostile Shipping:** Navy will have to deny any adversary from using its maritime forces against the territory of India or against Indian seaborne trade.
- **Search and Rescue:** Help in civil contingencies.
- **Constabulary Operations:** Anti-piracy and peace keeping functions.

Indian Air Force (IAF)

In war, the Air Force will make the decisive contribution in prosecuting operations successfully. The unique ability of the air force to concentrate military power when and wherever needed has become the dominant feature of modern warfare. It is primarily offensive in nature and is inherently a strategic force. The characteristics of the air power can be described as flexibility to adapt to changed circumstances, mobility to apply military power with speed and at long distances, concentration of power in time and space and the ability to carry out offensive action deep inside enemy territory. But IAF will be seriously handicapped unless they procure the force multipliers like airborne warning and control systems (AWACs), air-to-air refuellers, and Unmanned Combat Aerial Vehicles. (UCAV)

The limitations of air power are its sustainability in long drawn operations, dependency on bases and finite payloads and ranges. All these factors are equally applicable to Indian Air Force like any other air forces of the world. Therefore, considering the threats of the future, Indian Air Force will have its roles and missions tailored accordingly.

Roles of Indian Air Force:

- Defence of the nation against attacks through the medium of air and space.
- Control of air and space, to the required degree, to permit freedom of operations to own forces including Army, Navy, Coast Guard and relevant civil sectors.
- Operation in mutual support with army and navy total control of national airspace in both war and peace.
- Strategic Nuclear Deterrence. IAF is well suited for providing immediate retaliatory response in the event of a nuclear attack till such time the strategic nuclear forces are built up around Agni and Prithvi class of missiles. The response can be directed deep inside enemy territory.

Missions Related to Chinese Threat:

- In case of any future land battle in the hills of Himalayas and considering the limitations of air war in the hills, the present force structure of IAF will fall short to carry out any meaningful strikes to gain favourable air situation or interdiction missions with an aim to isolate Tibet from the mainland China. The induction of SU 30s may be a step in the right direction.

- On the other hand IAF should be ready to face the future offensives by SU 27s and SU 30s of PLAAF. Chinese Air Force is assiduously building up its long-range air war capability supported by air-to-air refueling and AWACs.

Pakistan: Considering the present asymmetry of force level vis a vis Pakistan, IAF should be in a position to carry out classical application of air power to gain positive advantages which would give the ground forces freedom of movement. But unless and until this asymmetry is maintained by future procurement and force updation etc., IAF would soon lose the advantage. Induction of AWACs, air-to-air refueling capabilities and SU30s etc., in good numbers would be steps in the right directions. It should not be forgotten that " Air power today has reached the status of the dominant factor bordering on a pervasive impact on war. It is true that air power may not win a war by itself alone; but without it no major war can be won[5] ."

Island Territories: IAF does not have any permanent presence in any of its Island territories except a token presence on Car Nicobar Island. Therefore, IAF will have to be ready for quick deployment of forces in the Island territories wherever possible and to mount operations from mainland as and when called for.

Conclusion

Armed forces of any country should be fundamentally created, organised, trained to perform combat roles to preserve the national security. Its main purpose is to kill people most efficiently and defeat any aggression across the border. The next question arises, should the armed forces also carry out **nontraditional**

roles as well? If so, what should be composition of the forces? Huntington while spelling out this stated, " All such roles should be spillover uses of the armed forces, which can be performed because the services possess the organisation, training and equipment that are only maintained to defend nation[6]."

The Indian armed forces have been carrying out these traditional and non-traditional roles quite effectively since Indian independence. It has fought four wars and continuously been fighting insurgency and cross-border terrorism. And if analysed correctly it may be seen that in each year of the last 60 years, the armed forces had been called out for providing assistance to the civil authorities continuously. The contributions of Indian armed forces to all UN Peacekeeping effort have been noteworthy.

It is very interesting to note that India has borders with seven neighbours but none of them are particularly favourably disposed to it, particularly China and Pakistan. Other smaller neighbours are only vocal about their endless hostility to this country for various reasons of their own but cannot afford to become openly belligerent due to inherent disadvantages. But Pakistan and China who are also declared nuclear powers, had never accepted Indian position in the subcontinent and committed acts of hostility whenever there was an opportunity. The situation is not going to change in foreseeable future and India will have to live with this hostility for many years to come. Roles and missions of the Indian armed forces have to be tailored to fit into such realities and sufficient peacetime preparations should be carried out to prevent Pakistan, China or any other nation acting against our national security interest.

Notes

[1] Daniel T Kuehl and Charles E Miller, "Roles and Missions, and Functions: Terms Debate," *Joint Forces Quarterly*, Summer 1994, p105.

[2] Daniel T Kuehl and Charles E Miller, "Roles and Missions, and Functions: Terms Debate," *Joint Forces Quarterly*, Summer 1994, p103.

[3] Department of Defence Directive 5100.1, "Functions of the Department of Defense and its Major Components," September 25,1987, pp19-21.

[4] Samuel P Huntington, "New Contingencies, Old Roles," *Joint Forces Quarterly, Autumn 1993,* pp38-39.

[5] Air Commodore Jasjit Singh, "Air Power in Modern Warfare," Lancer International presentation, New Delhi, 1985, p21.

[6] Ibid. p43.

5

Unmanned Fighters May Replace Fighter Pilots

It is likely that some form of air power will be the principal driving force before the surface forces are able to make contact with the enemy in major battles of the future. It is more or less definite that the Unmanned Combat Aerial Vehicles (UCAVs) will play a definitive role towards augmentation of air power. These will be smaller, faster and more versatile than its manned counterpart. The futuristic unmanned combat aerial vehicles will be able to penetrate airspace that may be considered dangerous for manned aircraft. These could carry out reconnaissance, surveillance, or could be used for jamming enemy air defence system. This has become a necessity in the face of the quick reaction, low looking missiles that are deployed by almost every warring nation. unmanned aerial vehicles (UAVs) made their presence felt in the Gulf war and more recently in the Kosovo war. The Israeli UAVs in the Bekka valley war[1], caused havoc to the Syrian Air Force.

In his key note address in the Aero India show in Bangalore in February 2001, then Chief of the Air Staff, Air Chief Marshal Typnis, laid great stress on the UCAVs and mentioned that "UCAV has not only the attraction of keeping combat casualties low, but overcomes human limiting stipulations on "g" crunching manoeuvres, and stringent cockpit environment for protection and survival of the pilot"[2]. No doubt this shows the keen interest evinced by the Indian Air Force to induct these futuristic technology to recast its force structure.

There is a volume in the Jane's series titled *World Unmanned Aircraft* by Kenneth Munson[3], which describes the programmes from 1988 till date. This may be the most comprehensive and contemporary source data to know about the UCAVs of the world. Despite some serious effort all over the world, only a few of UCAV programmes can be called as true success stories. Many have withered from lack of interest and resources and a few have been a spectacular failure. UCAVs fly more complex missions than cruise missiles and are more autonomous than remotely piloted vehicles. The distinguishing characteristics are the automation of complex functions. The main requirement of these unmanned vehicles is to perform tasks as part of the main systems and return unharmed as in the case of a manned aircraft. The operator or the remote pilot will be guiding the vehicle through its mission and bring it back sitting at a location as the mission demand. But it must be understood that the UCAVs are not a panacea. Some missions can benefit by the use of UCAVs but some other missions may be left to the manned flight. It is for the Air Force to determine the correct mix of manned and unmanned aircraft in future battle scenario.

The decision for manned versus unmanned aircraft application requires a careful consideration of the following factors:

(a) The scenarios to be encountered.

(b) The missions and tasks to be performed.

(c) The alternatives.

(d) The relative risks.

(e) The relative costs of the task.

(f) Availability of the technology.

Need for UCAV

The time for acceptance of UCAV appears to be here for a number of reasons. Declining force structure and equipment, necessitates innovative thinking about more cost effective solutions for the air forces. The air forces of the developing nations always find themselves at a disadvantage to fill the vacuum between technology and the force level. Critical condition of the Indian Air Force is well known. It is facing obsolescence in a few years time and there is no feasible cost effective solution. Replacement of the ageing Mig 21s is not possible by procuring off the shelf procurement of state-of-the art aircraft. Even if they were available, the cost of procurement would be prohibitive. Present defence expenditure cannot support such replacements of the manned fighter. On the other hand, the pilot training is getting increasingly costly and there are not many from the younger generation who would like to join the fighter force in preference to a highly paid managerial job. The factor of increasing rate of accidents is also causing a serious worry in the air force hierarchy. That is why there are opportunities for UCAVs to replace some of the manned missions. These could provide lower operational costs, and increase sortie rates.

Therefore, we need to assess the concept as well as technologies of platforms, weapons, and human factors as they will contribute directly to accomplish air force tasks. But it must be noted, that, UCAVs are not a complete solution to air force problems. These will never be able to replace the manned missions completely. The force mix decision process is especially complex for unmanned vehicle, because, the introduction of a radically new weapon system carries a great deal of uncertainty.

It is expected that the experiences gained form the UAV operations will assist in weaponising them to become a potent air power component.

UCAVs no doubt will give a different dimension to the concept of air power and its applications. Concept of weaponising UCAV may seem a radical idea but if this were examined closely the apparent reasons would come out very clearly. As a result a number of nations have become interested to weaponise the

UAVs in preference to manned fighters. The reasons are very clear. UCAV platform, sensor and weapons technologies have matured sufficiently to permit low risk, and quick development to weaponise UAV. The operational risks may be seen from different angles. Firstly, utility of this unknown weapon system may create control and coordination problem depending upon the role it is employed. But the greatest advantage that will accrue to a air power proponent is that the UCAV could be pressed into service with no fear of any loss of life. Suppression of enemy defence would not be an easy task in the future air war scenario but the UCAV could play some decisive role where the manned flight will face unacceptable losses. Therefore the air power protagonists will seek certain qualities in this new kind of weapon system[4]. These are:

(a) **Endurance:** UCAVs should have much greater endurance than any manned flight, because the question of crew fatigue will not be there. This will give the ability for persistent surveillance, continuous deterrence, reduce the number of aircraft to carry out the same job and will reduce cost of coverage of areas.

(b) **Operating Ability**: These should be able to carry weapons and perform high attrition combat tasks and operate in contaminated area, because the future wars may see the use of chemical and biological weapons. UCAVs can be used in provocative roles if a situation arises. If need be, these machines can be sacrificed in the interest of a mission. But it should not be forgotten that operational deployment would have the greatest disadvantage of not having the human reasoning power, as such the command and control problems should be resolved before its operational deployment.

(c) **Automated:** UCAVs are simpler and easier to operate and that is why cost of training is much less than its manned counterpart. And just because the manned intervention in the air is not necessary no crew safety training or systems are to be integrated. Control surfaces can be made simpler which will bring down the cost and reduce complication of the design. Crew will be operating the UAV/UCAVs from a remotely based control room in a controlled environment as such the crew fatigue will be very little. If long hours of watch are required then the crew will be able to hand over control easily. This is unthinkable in case of manned fighters.

(d) **Quick Response**: It should be having the capability of quick response within the zone of operation. These should be able to operate behind the enemy lines; combined attack with multiple weapons should be possible. Broad area can be covered with multiple sensors, persistent surveillance, reduced system vulnerability.

(e) **Capabilities for High Altitude Operations**: Survivability, performance enhancements, broad area coverage, better viewing angle.

(f) **Low Altitude Operation**: Survivability, short range of operations in role of jammer, threat from smaller weapons.

Technological aspects expected in UCAV[5] s are:

(a) Low lifecycle cost.

(b) Engines should be fuel efficient for better range.

(c) Simulation based design for systems relevant to humans.

(d) Lower signatures will assure lower observability.

(e) Mission planning should be automated, flexible, fast, utilising parallel computers.

(f) On board processors should have excellent performance and cost should be low.

(g) Sensors must be modular, lightweight, UAV tailored.

(h) System design to be based on robust system with very low rate of failure.

(i) Integrated design/simulation, manufacturing automation.

(j) Crew selected and trained using modern methods.

(k) Vehicle management system to be optimised for UCAVs in relation to performance, weight, cost and automation.

(l) Structure of the UAV should be tailored composite with very low part count and high fuel fraction.

(m) Weapons must be small, modular and be an integrated system.

Operational Missions[6]

Attacking Fixed Targets: UCAVs could be employed to attack high value heavily defended fixed targets. Use of manned aircraft would cause unacceptable losses to man and machines. Also these could carry out supporting operations in the missions of strategic attack, interdiction and close air support. Given the location and type of target and desired weapons effects, a target attack mission would determine tactics for target acquisition, weapons effects, collateral damage and terminal system based on GPS, electro optical, IR or millimeter wave.

Attack Moving Targets: Typical missions covered by moving targets attacks are interdiction, strategic attack and close air support. This could be made simpler because the UAVs would be capable of loitering for a long time át high or low altitude carrying sensors to detect and cue loitering weapons platforms to attack identified targets. The weapon platforms in this case would mostly be manned fighters. But in future, UCAVs could also be applied.

Jamming: Application of UCAV in role of a jammer could be most effective. These platforms could operate at high altitude, long endurance/low observable electronic support measures/ electronic countermeasures platform supporting multiple strike/ bomber attack operations in standoff or close orbits. Long endurance would permit the UAV to support multiple strike or bomber attack on a heavily defended target. The vehicle would have the ability for preplanned orbit navigation or ground and airborne dynamic re-tasking in support of revised targeting. The concept of jammers preceding strike force is being practiced by many air forces to deliver weapon load on targets defended by quick reaction missiles. The UAVs if matched well with the strike force could be used to neutralise enemy fire control and tracking radar. A decoy UAV can be deployed before the actual strike force approach the target. These could replicate the radar or IR signatures.

Suppression of Enemy Air Defence (SEAD): UCAVs could be deployed to detect enemy air defence systems and pass the detection and precision location data to elements of the SEAD network that would deploy attack weapons. The best way to deploy UAV in present scenario would be to provide a good safe loiter time so as to pick up emitter data and pass it on to command post and in turn the manned fighter could be sent to deliver the actual weapon load. However, technically it is feasible to deploy UCAVs to carry out actual attack. Enemy air defense system could be saturated by the appropriate deployment of the UAVs.

Air-to-Air Combat: Air-to-air combats have mostly been restricted due to human limitations. The manned fighters could sustain certain "g" force for a specific period. Every fighter aircraft designed so far had this central theme as one of their main limiting factors. But the unmanned fighters will not have any such restrictions and as a result their maneuverability will be far superior to the manned fighters. Therefore, high "g", high-speed interceptions would be a definite possibility.

Surveillance/Reconnaissance[7]: UAVs have the potential to collect intelligence data throughout day and night from anywhere in the enemy territory regardless of weather and transmit it to the command post. UAVs have the definite capabilities to fly close to the target, long dwell and loitering and if it is added with low observable capabilities, these machines could overfly a heavily defended target.

Communication Nodes: Multiband multimode communication relay and switching nodes will contribute to the force enhancement role of the three Services. The value of such support will manifest in every offensive operations when a tactical communication network is limited in keeping pace with the fast moving forces, not only in physical speed but in power, frequency, bandwidth, available channels and avoidance of interface. Considering Indian terrain and mostly in hills the forces widely dispersed and beyond line-of sight communications will benefit the most by the application of the UAVs[8].

Operational Analysis[9]

Major air powers of the world understand that UCAVs are coming of age with mature technology and tailored roles to perform; despite setbacks in many fields. Many people on the other hand, wonder why these are not more widely used. Somebody talks about a pro pilot bias. Some people have different perception; they say that unmanned vehicles are incapable of performing the mission like a manned aircraft since these are unable to respond to the combat environment's dynamic changes. This is assumed that UCAVs are no more intelligent than cruise and ballistic missiles. It does not take into account the possibility of a real or part time operator or the quality of feedback provided by the UCAV sensors. Unlike non-lethal UCAVs, unmanned system carrying lethal munitions could have devastating effects in an accident or system-related malfunctions. There are also opinions which directly challenge UCAVs capability of performing traditional manned aircraft missions and that could be a direct threat to an air force as an institution. At certain quarters it is argued that air force has accepted the induction of unmanned aerial vehicles when faced with other services infringement on traditional air force missions. This is also true that today accident or mistake justification does not warrant death or destruction. Society has become more sensitive to death and destruction as the information age provides real-time world event reporting. In addition, technology has legitimised precision warfare and criminalised collateral death and destruction resulting from the use of lethal force. An autonomous system has no way of knowing how the situation may have changed since it received its mission orders. Biggest problem will be faced once an internal disorder takes place. The UCAV will not be able to detect the fault and could create certain dangerous situation.

These may be the reasons why air forces would be facing such complex questions. How many UCAVs to buy and for what missions? Such analysis would indicate certain route to follow. In broader term, it should be defined what the air force wants to accomplish by adding UCAVs to its force structure. Is it to supplement manned missions and reduce training cost? Is it to save money? Or is it to do a task or mission, which is dangerous for manned missions? Each of these questions has to be carefully studied and a viability of the UCAVs has to be established.

Careful assessment has to be made for each manned flight and critical assessment should be carried out, so that UCAVs can be employed appropriately to replace manned flight in each of the identified missions. It is, however, not necessary that the replacement has to be on one-to-one basis. A manned flight if necessary can be carried out by many UCAV flights provided the time permits. Ultimate decider would be that, UCAVs perform missions more cost effectively than a manned flight and that too in a very hostile environment. Force structuring analysis is needed by the air force to evaluate the effectiveness of UCAV. Detailed simulation of performances of all systems and subsystem would be appropriate. Any mission level model with application of both manned and unmanned flights could be compared before the UCAVs are inducted in actual operations. This would also be helpful to assess survivability rate of the UCAVs[10].

Functioning of Command and Control[11]: Positive control of UCAV/UAVs would require the important C3I functions, including the human operators to intervene quickly to regain control of an errant autonomously controlled vehicle. And this is way the UCAVs can be integrated successfully into air operations. The need is to interface UCAV capabilities with the existing and emerging infrastructure for command and control system. Each mission will have its own needs, design consideration, onboard sensors, computers and weapon systems.

New ideas in autonomous controllers and associate systems that support and collaborate with human operators in a hierarchical command structure and new concepts for passing targeting and intelligence data from the sensors to shooters are to be addressed for proper utilisation of this new but effective tool to wage future air war. It has to be found out whether the present or emerging C3I architecture are adequate for UCAVs induction into the force structure for the air force. The main challenge in the realm of command and control for these machines would be positive control in shared air space with manned flights. The technology requirement would be powerful software and hardware to sustain real time onboard mission planning for the complex set of UAV missions in any future conflict dominated by air power.

Survivability of UCAVs[12]: Survivability of UCAV is a critical issue. Certain losses of this low cost technology may be acceptable in comparison to its manned brethren. But loss of cost effectiveness may not be acceptable to the operational community. Therefore, survivability features must be brought in the case of each specific UAV design. This may have to be balanced carefully between cost, mission requirement, lifecycle cost and maintainability. UCAV design has to take into account two different dimensions of threat.

- **High Density of Threat**: Where manned mission could be a futile exercise but certain amount of losses may be acceptable.
- **Low Threat**: On the other end of the threat spectrum would be little or no threat environment. Longer endurance, an inherent capability of all UAV operations could make it less survival, but multiple numbers of these lesser-known air assets may compensate to an extent.

All UCAV design or procurement will require the appropriate mix of signature control, tactics, emission control and onboard countermeasures. To this respect the design consideration has to take the route of any manned aircraft. Threat avoidance may be the central issue of all manned or unmanned missions. That is why, UCAV mission also should be planned carefully giving appropriate priorities for all factors for effective air power planning. Manned and unmanned mix for any mission will entail new mission planning system not applied earlier in any air conflict fought in recent history. Kosovo war could have been cited as an example of the application but the air space of Kosovo was not as hostile as it could have been.

Unmanned Aerial Vehicles

UCAVs are on the drawing board and it is expected that some of its lesser-known varieties will make their appearance sooner than predicted. Presently usable load carrying capacity of the front line UAVs is limited and there has been an attempt by the Americans to fire Hell fire missiles from their Predators (UAV) during Afghan air war. Indications are very clear that the development of UCAV will piggyback its ride on the UAVs learning from their recent operational applications. During the war of Kosovo, the coalition forces deployed a large number of UAVs and these unmanned aerial vehicles played a significant role by sending battlefield images back to the allied command post and targeting cell. The entire air war depended on remote sensors and long distance Precision Guided munitions to keep the pilot attrition to the minimum. The result of the air campaign was devastating. Following are the UAVs deployed by the allied forces in Kosovo:

- Crelle (France)
- Bombardier (Germany)
- Meteor (Italy)
- Phoenix (USA)
- Predator (USA)
- Hunter (USA)

Most recently, the USA deployed Predators (UAV) for recce and surveillance prior to commencement of aerial bombardment of Afghanistan and lost two of them. The presence of these UAVs and their firing of Hell fire anti-tank missiles is the coming of age of an UCAV. Global Hawks were also pressed into service for reconnaissance and surveillance during the Afghan conflict.

UAVs of Pakistan[13]: Pakistan's air weapons complex showed two new tactical UAVs at the IDEX exhibition in Abu Dhabi in April 2001 the Bravo and the slightly larger Vision-1. Both are of typical pod and twin boom pusher configuration. Bravo entered production some 18 months earlier and is currently used by the Pakistan Army for border patrol and low enforcements. Vision-1 is being prototyped for planned production in 2002, while long range Vision-II is under development.

Chinese UAVs: China has shown keen interest to develop her own UAVs. Following are some of the Chinese UAVs:

- BUAA Chang Hong[14] (Long Rainbow) may be the copy of US Teledyne Ryan Model 147H. Two prototypes were built in 1972. The UAV entered service in 1981 for recce purposes with a wing span of 9.76 mtrs, mission payload 65 kgs, maximum (max) launch weight 1,700 kgs and endurance 3 hours.
- Xian ASN-104 and ASN-105: These are reconnaissance and surveillance UAVs developed by Chinese Pilotless Vehicle Research Institute. Out of the eleven types seven received the Governmental approval. ASN 105 has slightly better trange than ASN 104. The UAV has a wing span of 4.30 mtrs, max payload 30 kgs, max launching weight 140 kgs and endurance 2 hours.
- XIAN 206 Multirole UAV: First time displayed at China Air Show in November 1996. This is launched by booster rocket and recovered by parachute. It has a wing span 6.00 mtrs, max payload 50 kgs and endurance (max) 8 hrs.

UAVs of India

India also went ahead to develop some variants of UAVs. These are:

- Kapothaka[15] (Sparrow): High wing monoplane, with radio command link via UHF uplink. It is launched by rocket booster and recovery is by two stage parachute. Link loss routine consists of a slow climb and recovery automatically. Wing span 4.50 mtrs, max payload 20 kgs and max endurance 1hr 30 mts.
- Nishant:[16] Developed in 1990s by the ADE. It made its first public appearance in December 1996. Initial production has been entrusted to Taneja aerospace. It will have the capability of real-time day/night sensors. It will also carry electro-optic visible IR sensor, laser designator and mini panchromatic film camera. It will have a wing span 6.50 mtrs, max payload 60 kgs, endurance 4 hrs and ceiling 4,000 mtrs. It is expected to enter service with Indian Army soon.

UAVs of Indian Air Force: In his keynote address at the international seminar on *"Aerospace Technologies: Development and Strategies,"* Aero India, Bangalore, February 2001, Air Chief Marshal AY Tipnis pointed out that:

> " Uninhabiting the aerial platform is already underway for a variety of applications. While the uninhabited combat air vehicles (UCAV) are in my reckoning, still at least two decades away, the concept has already taken a firm hold. The stand-alone UCAV is not likely to be progressed directly but will possibly evolve as a directed buddy to other manned aircraft. The UCAV has not only the attraction of keeping combat casualties low, but overcomes human limiting stipulations on G crunching maneuvers, and stringent cockpit environment for protection and survival of the pilot. This will give it a mind stunning performance, with 30g yanked turns at astounding speeds of 15-20 mach. And all that survivability will be high for ensuring mission accomplishment; high survivability will also give it extraordinary resilience, increasing the reusability factor manifold[17]."

Indian Air Force has procured both Heron and Searcher UAVs produced in Israel. Details of both Heron[18] and Searcher II[19] are available in Jane's *Unmanned Aerial Vehicles and Targets* edited by Kenneth Munson.

Airspace Management

The issue of airspace management is the key to successful operation in civil and military environments. The UAVs that Indian Air Force procured or could develop will have to operate in diverse airspace environments, which would require appropriate airspace management. It is a relatively long climb to uncontrolled airspace for the high altitude long endurance aircraft. Such climbs would require long climb corridors through controlled and at times crowded airspace. Unless these are carefully planned and monitored there could be disasters.

In Director General Civil Aviation (DGCA) controlled air space, a separate set of rules are required to be formulated for easy operation of these flying machines. The rule of 'see and be seen' can be the best solution. That will involve a chase aircraft or the use of restricted airspace for all UAV operations. Or else Indian Air Force can obtain a one-time clearance for such operations to outline the desired approach to UAV flight operations.

Lower altitude UAV operations will pose similar peculiarities that of attack aircraft concepts, which would involve flight through controlled airspace environments even for training and exercise missions in peacetime. Additional precautions would be necessary in wartime, when airspace environments would be extremely crowded in certain areas. In India, UAV operations factors are generally unknown. As Indian Air Force has acquired UAVs, they will need to develop rules, procedures and proper planning for utilisation of these assets.

Air space management in the tactical battle area (TBA) may have been the most frustrating issue for the Army and Air Forces of any nation. In Indian context, both the manned flight and UAVs operated by Army and Air Force will share the skies of

TBA. Indian Army will be fielding the DRDO built Nishant and Searcher I, whereas, the Indian Air Force will have their Herons and Searcher II. Though the UAVs will mostly fly at higher altitudes for recce and surveillance, yet, for laser designation these will have to come down to a lower level.

For all military operations, airspace management in TBA is the responsibility of the Air Defence Direction Centre (ADDC) or Joint Air Defence Direction Centre (JADC) as the case may be. There are laid down procedures for airspace coordination. The ADDC/JADC decrees the sole use corridors, designates control authorities (Airborne Warning and Control (AWAC), Control and Reporting Centre (CRC), establishes rules, and provides procedures for the safe passage and orbits of all manned aircraft, long range artillery, air defense weapons and missiles. Free fire zone, and flight corridors are established as a function of time of the day and hence a highly dynamic airspace deconfliction process is essential. It is correct for Indian Air Force to begin to think through the issue of airspace deconfliction for the broad range of environments and scenarios expected in the future battlefields in the subcontinent. Regular exercises and training with these UAVs could resolve many issues.

It is now abundantly clear that the senior leadership of Air Force has displayed the vision to introduce an unknown system to be employed in autonomous and complementary roles. Therefore an aggressive effort will be needed to introduce UAVs to the fighting forces and to integrate the capability into the joint planning structure for the future wars in the subcontinent. A strong vision, development of operational concepts should include UAVs as close complements to manned aircraft. Every effort will be required to clearly define and integrate mission system, weapons, human systems, communications connectivity, mission planning and self protection.

Offensive and Defensive Air Operations: UAVs are expected to be deployed in surveillance and intelligence gathering. Considering their endurance these are expected to be loitering at the target area for a long time, which will enhance their vulnerability to enemy air actions. Therefore, for all offensive deployment, Air Force will have to make sure their survivability and recovery. Generally, due to its size, UAV presents a very small radar signature and there are not many types of radar in the subcontinent, which could pick up the echo and direct the airborne manned fighters to make an effective kill. For a pilot in the air, it would be necessary either to pick up the UAV on his radar screen or have visual contact to fire any kind of weapon. Both these process will be seriously affected if the airborne radar does not detect the target and all aspect heat-seeking missile fails to home on to the target. It is not known whether visual contact is possible under ground radar control and carry out effective interception of these lesser known aviation assets of the future, which are going to make their presence felt in large numbers in any future conflict in the subcontinent. It is also not known whether these tiny but potent systems could be picked up by the SAM 6 or SAM 8 class missile guidance radars These important aspects will have to be taken into account to counter the Pak UAV threat and effectively utilise own UAV assets.

Conclusion

Experiences of the Gulf and Kosovo wars brought out several strategies to the forefront, and supremacy of air power had been clearly established. Air power made it easy for the ground forces to complete the annihilation of the Iraqi Army during Gulf war. War in Kosovo was brought to a logical end by the application of air power with no or marginal involvement of the ground forces. That is why the war strategists do not have any doubt that, some form of air power will decide the course of the battles in future. Such deliberations naturally bring focus on the factor of UCAV/UAV. These could be add-on forces to augment future air wars. These would be smaller, faster and more agile than manned fighters. Such unmanned vehicles could be used in preference to manned flights in a dense air defence battle space, optimised by low looking radars and QRMs (quick reaction missiles).

UAV concept has been proved in the current air wars. And there are indications that the UAVs are gradually getting equipped with weapons system to fight future air battles traditionally fought by manned fighters. These could meet the operational need and fit into a sound plan of operations. Historical biases for manned aircraft should not be the stumbling block for embracing this technology in the futuristic technology driven wars. A number of countries including Pakistan have taken steps to add this new dimension of air power to their arsenal. Indian Air Force has taken the procurement route to augment its firepower. But a coordinated attempt has to be made by DRDO and the Services to bridge the technological gap at the earliest and induct own UAVs with a final aim to have our own UCAVs.

These developments point out to the fact that, future air wars in the subcontinent will witness a different kind of air operations due to a large-scale induction of UAVs. Offensive and defensive operations against these airborne elements have to be understood and precautions to be taken. Also problems that will arise towards air space management in the TBA will have to be resolved, before the next application of air power.

Notes:

[1] Air Vice Marshal RA Mason, "War in the Third Dimension" in Air Marshal Sir Michael Armitage ed., *Manned and Unmanned Aircraft,* London, Brassey's Defence Publishers, 1986, p193.

[2] Air Chief Marshal AY Tipnis, Keynote Address at Aero India Show at Bangalore in February 2001 (www.discovery.com)

[3] Kenneth Munson, *Jane's Unmanned Aerial Vehicle and Targets,* Jane's Information Group Ltd, Surrey, UK, Nov 1998.

[4] <www.raf.mod.uk/airpower/doctrine>

[5] < www.webruler.com/aircombat/drone_war.htm>

[6] <www.au.af.mil/au/database/projects>

[7] Bill Sweetman, *Hale Storms to New Height,* International Defence Review, March 2001, pp50-55.

[8] Roger W. Gallington, Harry Berman and John Entzminger, *Unmanned Aerial Vehicles,* Science Application International Corporation (Kent, Washington 1998), Defence Advanced Research Projects Agency(Arlington, Virginia 1998).

[9] n1 pp187-202.
[10] Ibid. p274.
[11] Edward Waltz, *Information Warfare Principles and Operations,* London, Artech House, 1998. pp268-269.
[12] n8, pp288.
[13] 15 News Brief, Jane's Information Group Ltd 2001, 16/06/2001.
[14] Kenneth Munson *Jane's Unmanned Aerial Vehicles and Targets,* April 1999, p15.
[15] Ibid pp59-60.
[16] Ibid p61.
[17] Excerpts of the Keynote address of Air Chief Marshal AY Tipnis at Aero India Show at Bangalore Feburary 2001<www.discovery-.com>
[18] Kenneth Munson, *Jane's Unmanned Aerial Vehicle and Targets,* Jane's Information Group Ltd (Surrey, UK, Nov 1998,) JUAVT-ISSUE-O.
"Searcher MKII, Heron Pchela –1T": Heron UAV system is an operational fourth generation long-endurance medium altitude system and is based on leading edge technology. It provides deep penetration, wide area, real time intelligence to national agencies, theatre commanders. The Heron has modular space up to 250 kgs for customer furnished equipment is interoperable with other Israeli developed UAV systems. Heron can have SAR, Maritime patrol radar, Satellite data link capability. **Operational Capabilities.** In operational modes Heron has real-time payload and UAV control; GPS based interruptible airborne mission controller; full autonomous flight and autonomous return on data link loss. **Datalinks.** It has direct line of sight data link; UAV airborne data relay for beyond line of sight data link; dual real-time command uplink and single real-time data and voice downlink. **Launch and Recovery.** Automatic take off and landing. **Endurance.** 40 hours. **Range.** 200-350 km may go upto 1000 km. **Altitude.** 30,000ft
[19] Ibid. **Searcher MK II:** Searcher Mk II is the tactical multi-payload UAV system. This can be used for surveillance, reconnaissance, target acquisition and artillery fire direction. Standard payload can be TV and IR combination. **Operational Mode.** Searcher II has real-time payloads and UAV control mode. This has GPS based airborne mission controller mode with real-time manual interruption capability. Also this is equipped with autonomous return mode activation on data link loss. **Data Links.** Direct line of sight data link. UAV airborne data relay for beyond line of sight data link, Dual real-time command uplink and single real-time data and video downlink. **Launch and Recovery.** Automatic take-off and landing. **Endurance.** 15 hours with range between 200 to 250 kms. **Altitude.** 20,000ft. **Max take off weight.** 426 kgs. **Dimensions.** Length 5.85 mtrs **Wing span** 8.55 mtrs.

6

Emp Weapons:

The Force Multipliers

During the Kosovo war a Tass wire story reported for the first time about the presence of a new weapon being tested by the USA. The weapon could have been fabricated at Los Alamos and this was the most secret weapon of USA designed to destroy radio electronic equipment. The Russians also went on to report that these bombs were dropped by B-2 stealth bombers to generate electromagnetic pulse (EMP). Pentagon officials however denied its use and reported that there was no intention to use any such weapon in the Yugoslav conflict. But the assertion that the U.S had indeed used such a weapon was made during the first few days of the Kosovo conflict which came direct from the Russian Defence Minister. He went on record to inform that the USA was using Yugoslavia as a test range for its latest secret means of destruction[1]. The USA while categorically rejecting the report went to state that they wanted to keep the capabilities of the weapon under complete secrecy and would use the same in a future campaign against a sophisticated foe. Existence of the weapon had not been denied.

It seems that the weapon can be delivered by cruise missile or by any other stand off platform and can be exploded within yards of a target. The explosion would be followed by generation of high power microwaves that can disable electronic circuitry in computers and communication equipment. This EMP could erase software and computers. David Fulghum[2] while reporting on this unique bomb stated that on a clear day the effect could be felt in an area as large as a football field. Col John Alexander has written in his book *Future Wars*[3] about the use of High Power Microwave (HPM) or electromagnetic pulse weapons (EMPs) and carbon graphite fiber or ribbon bombs. While the effectiveness of carbon–graphite bombs is now well established from raids on Baghdad and Belgrade, the HPM weapons are less known. Alexander thinks that the concept of EMP bomb may be comparable to a lightning strike in the vicinity of computers or other electronic equipment.

The electromagnetic pulse effect was first observed during the early testing of high altitude airburst nuclear weapons. EMP is caused by the rapid release of gamma radiation from the nuclear explosion. The release of these particles at the speed of light will produce regions of positive and negative charges as atmospheric molecules are stripped of their electrons. These charges will propagate through the air at the speed of light and can have significant effects on all electromagnetic signals within line of sight of the nuclear detonation. The electromagnetic pulse is in fact an electromagnetic shock wave. The EMP energy produces such a storm of electromagnetic field that it produces short lived transient voltage of thousands of volts (Kilovolts) on exposed electrical conductors like wire or conductive tracks on printed circuits boards where exposed.

One significant factor in EMP effect is the amount of coverage desired. The area of exposure depends upon the size of the yield. It was also observed that high altitude electromagnetic pulse became the highest concern as the entire electromagnetic spectrum could get affected. It was also found out that high altitude burst could produce large amplitude EMP field over thousands of kilometers. Peak energy fields can reach levels of 50 kilovolts per meter. The peak levels can be reached very quickly and will have large broadband frequency coverage extending from Direct Current to 100 Mhz frequencies[4].

The covert testing of the weapon in an active field of Yugoslavia has demonstrated that the EMP bomb is round the corner[5] and that the US will not hesitate to use the same in any future conflict considering its negligible collateral damage making capability. The idea would be to immobilise the enemy before destroying him. Development of this type of weapon has more or less become a necessity because of the changed environment. Information flow has to be strengthened for own side and it has to be stopped for the enemy. This could be the cleanest bomb, a clean match winner. No doubt this is a future requirement.

While carrying out spectrum comparison it was determined that EMP spreads in a waveform. Diagrammatic representation of the same is shown below:

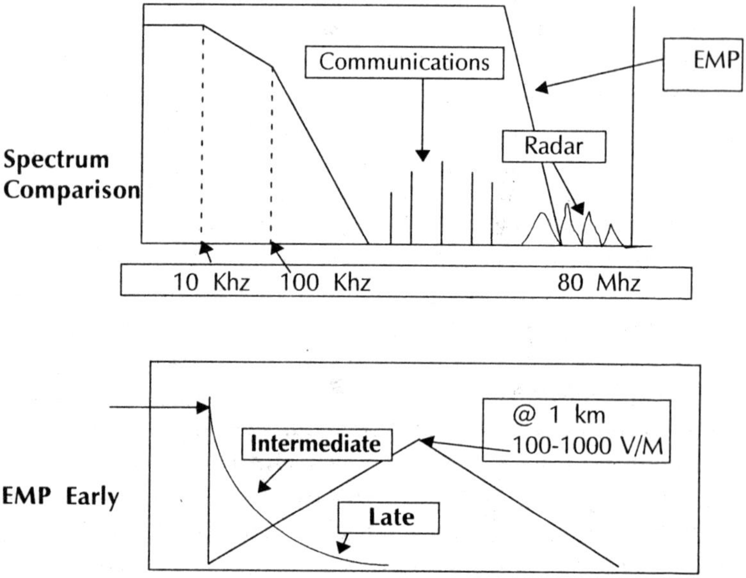

Fig 1. **Electrical Field Environment**

This waveform has been broken down into three segments[6]. The first is called early time EMP, and is the most devastating segment of the waveform; maximum levels of energy are produced in a very short time. Because of the intensity of the energy and the speed of the waveform, unprotected circuitry will be damaged or destroyed. On the other hand late time will occur at about one second after generation and can last up to 1000 seconds. During late time EMP low levels of energy are induced into the varying magnetic fields in the earth, which results in electrical fields being determined by the earth's surface resistance. This energy can pose a threat to long landlines such as telephone, power and submarine cables.

Conventional Electronic Bombs

Timothy L. Thomas while discussing the desperate situation of the technologically antiquated nations stated that,

> "Rapid technological change presents a specific new challenge to strategists: the requirement to master the emerging forms and functions of information technologies. New developments in managing information can create suspicion–even paranoia—among nations that lack the enabling technologies we take for granted. Technologically antiquated nations, those without as well as those whose infrastructure are outdated, could be more inclined to preemptive behaviour when they perceive a threat than would those states more attuned to the capabilities and limitations of the latest technologies."[7]

In many areas, technology for designing an electronic bomb is quite mature. And it is estimated by Western experts that the fabrication of an effective Flux Compression Generator (FCG) can be accomplished with basic electrical materials common plastic explosives such as C-4 or semitex and readily available machine tools such as lathes and suitable mandrels for forming coil that would be more than sufficient to build a bomb. It is also estimated that the cost of building one bomb could be as low as $ 1000-$ 2000

Key technologies, which could be used in the near future, are discussed here:

- Explosively Pumped Flux Compression Generator (FCG).
- HPM devices based on Virtual Cathode Ray Oscillator or Vircator.
- Propellant driven magneto-hydrodynamic generators.

Flux Compression Generator

It was in the late fifties that the first demonstration of FCG was carried out at Los Alamos National Laboratory, and this could be the most matured technology today. The current produced by a large FCG could be 100 times greater than that produced by a lightning stroke. FCG device can generate energy to the tune of greater than mega joules in 100th of microsecond and produce peak power level exceeding tens of mega-amperes.

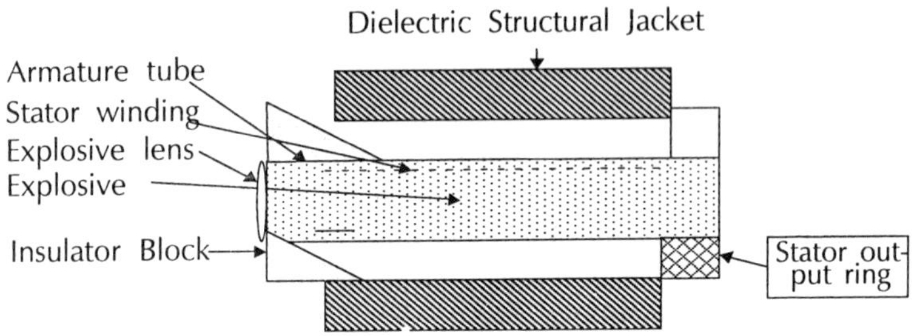

Fig 2. Explosively Pumped Coaxial Flux Compression Generator.

A cylindrical copper tube forms the armature of typical coaxial FCG as indicated above. The tube is filled with a high energy explosive. The armature is surrounded by a helical heavy copper coil, which forms the FCG stator. The stator winding can be split into segments with wires bifurcating at the boundaries of the segments. This could optimise the electromagnetic inductance of the armature coil. A structural jacket made of concrete or fibreglasses hardened with epoxy can be used to prevent the FCG from disintegrating at an early stage due to intense magnetic force. The weight of the bomb, supply of start current, and matching the FCG to the intended load carrier are the major technical issues of a FCG. When the start current peaks, the explosive is initiated, and explosion takes place in the armature, distorting it into a conical shape. Start current is isolated and stops the current inside. The magnetic field gets compressed, reducing the inductance of the start winding. This will produce a very high current pulse, which peaks before final disintegration of the device. FCG is generally constrained to the frequency band below 1 MHz. Many targets will be beyond this range of frequencies and as such will be difficult to attack[8].

Vircator

The Vircators could be the source of High Power Microwave source. Other devices that could produce HPM are Klystron, Magnetron, Reflex triode, etc. However, the best suited for the bomb would be the Vircators or Virtual Cathode-ray Oscillators. Vircator is capable of producing a very powerful single pulse of radiation and can operate over a relatively broad band of microwave frequencies. The basic idea of Vircator is to accelerate a high current electron beam against a mesh anode. Many electrons will pass through the anode and form a bubble of space charge behind the anode. If the space charge region is placed into a resonant cavity, which is tuned, correctly, very high power can be generated and microwave power can be extracted. Power level achieved by Vircator experiments ranges from 170 Kilo watts to 40 Giga watts with frequencies covering over the decimetric and centimetric bands. Output pulse duration of a Vircator is likely to be in the order of microseconds and is limited by anode melting.

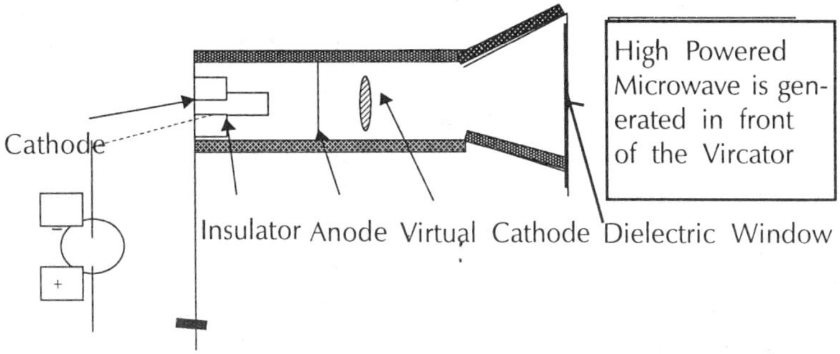

Fig 3 Axial Virtual Cathode Oscillator

Two types of Vircators —Axial Vircator and Transverse Vircator[9] are known to be used.

Magneto-Hydrodynamic Generators

Explosive and propellant driven Magneto-Hydrodynamic Generators are not too well known a technology. There is every indication that the range of development of this system is at a very elementary stage and thus precludes its being used in weapons systems. Best advantage of MHD generators are their compactness (liquid-metal MHD generators) and the absence of rotary parts. MHD can be used very easily as a compact source of electrical energy and that would be attractive for military applications.

Effect of Radiation

Present generation electronic devices like diodes, transistors, gate arrays and ICs are based on pure silicon slices. Their electrical properties depend upon the regularity and uniformity of the basic silicon crystal lattices. The initial total damage from Neutron radiation is proportional to the neutron influence, but there is a subsequent annealing process during which there is some degree of recovery. This apart, the damage could be permanent. Also it may be made clear here that it makes no difference whether the device is working equipment or kept on the shelf for future use. However, the annealing process will be longer in such cases.

Commercial computer equipment is particularly vulnerable to EMP effects. It is basically built with high-density Metal Oxide Semiconductors (MOS) devices, which are very sensitive to exposure to high voltage transients. MOS can sustain permanent damage with very little energy. Any voltage typically in excess of tens of Volts can produce an effect termed **gate breakdown**[10], which effectively destroys the device. Even if the pulse is not powerful enough to produce thermal damage, the power supply in the equipment will readily supply enough energy to complete the destructive process. Wounded devices may still function but their reliability will be seriously impaired. Similarly computers used in data processing systems, communications systems, displays, industrial control applications, including road and rail signaling in the advanced nations and those embedded in military equipment, such as signal processors, electronic flight control and digital engine control system are all potentially vulnerable to EMP effects. Degradation effect on few of the key elements employed in electromagnetic equipments are given below:

Light Emitting Diodes (LEDs): LEDs can suffer degradation in optical output by 10 to 20 per cent.

Photo Diodes and Photo Transistors: These are extremely sensitive to damage by Neutron radiation, and Phototransistors may suffer a loss in optical sensitivity.

Diodes: In case of diodes, there is a permanent increase in the forward voltage drop. This would require provision of additional heat sinking for high power devices.

Bipolar Transistors: Reduction in current gain and increase in collector emitter saturation voltage. This may point out about a requirement of high power high voltage transistor. But the design of the same may be difficult.

Lethality of Electromagnetic Weapons

The calculation of electromagnetic field strength achievable at a given radius for a given device is not difficult. But determining a kill probability for given targets under such conditions is complex and difficult because target types are very diverse in their design, frequency range and electromagnetic hardness. Moreover, certain types of equipments, which have been shielded against electromagnetic discharges of higher order, will withstand orders of magnitude greater than standard commercial equipments. Microwave can couple more easily than weapons of lower frequency range. That is why microwave weapons are more lethal.

Front Door Coupling: Let us take an example of an antenna subsystem, which is designed to couple power in and out of the equipment. This is the route for the power flow from the EMP weapon to enter the system and cause damage.

Backdoor Coupling: Similarly backdoor coupling takes place when electromagnetic field from a weapon produces spikes of large transient currents. These could be produced either by low frequency weapons or by electrical standing waves produced by a HPM (High Power Microwave) weapon. Equipment connected to electrical routings experience high voltage spikes and damage exposed semiconductors. As a result this could damage power supplies and communication systems and if the spike flows inside an equipment, it could cause serious damage.[11].

The damage causing scenario could be serious because low frequency weapon will couple into wiring infrastructure of the telephone lines, networking cables and power lines which generally follow streets, building contours and corridors. Deployment of the weapon will be of no consequence, because at least few segments of the wiring will be oriented in such a way that good coupling could take place by the emission of the electromagnetic energy. Electrical equipments like consumer electronic goods when exposed to low frequency and high voltage spikes similar to a lighting strike sustains extensive damage. Similar spikes would definitely damage computers and communication equipments (voltage as low as 50 V can inflict substantial damage to computers). The safe operating envelopes of some semiconductor devices are discussed here for easier understanding of the lethality of the electromagnetic shock waves. Breakdown voltage rating for Silicon high frequency bipolar transistors vary between 15 V and 65 V. These transistors are used in communication equipment. An essential component of computer is Dynamic Random Access Memories (DRAM), which are vulnerable to 7 Volts. Rating of CMOS logic is between 7 to 15 Volts, and microprocessors of the computer are vulnerable over a voltage of 3 to 5 V[12].

The above parameters clearly indicate that many of our equipments operating within the above voltage range can become an easy prey for the EMP weapons whether high yield or low yield. No doubt the damage will be extensive. Microwave weapon can substantially be more lethal than the low frequency weapon because it can bypass many protection devices built by the manufacture. Scientists are busy finding out the vulnerability vs. quantum yield of a weapon. Target diversity, individual electrical characteristics of the equipments' wiring and infrastructure around the target makes it difficult to predict the lethality of a weapon.

However, design factor should take into consideration the maximum lethality of an EMP bomb. To do that, it would be necessary to maximise power output and increase duration of radiation of the weapon. Enhancement of the known techniques like Flux Compressions Generation and Vircator effectiveness of High Power Microwave (HPM) may be the answer.

For efficient back door coupling it would be necessary to determine a known lethal voltage level, and that could be the basis to find the required field strength to generate voltage. The lethal radius for given weapons configuration can be calculated once the field strength is known.

A resultant several kilo volts meters is expected within the footprint of 400 to 500 mtrs of a Vircator type device producing 10 GW. 5 GHz HPM. Target to device distance can be calculated easily and safely estimated that the device can be exploded at a fairly longer distance from the target. This would be quite contrary to any conventional weapon.

Another design characteristic would be to maximise efficiency by making sure that the emitted energy is not wasted at the cost of lethality. This could be the researchers nightmare.

Targets for Electromagnetic Bombs

For a military man it will not be an easy task to identify targets for the electromagnetic bombs because of the complexity of the entire system. It may be easy to identify certain targets but to match the bomb capability with the target may not be that easy. Buildings housing various government departments and installations areas are easy to locate and target because of their geographical locations and can be attacked provided the aircraft can penetrate to weapon release point and height. Present generation smart bomb technology can be used to deliver E weapons to derive best possible results[13]. Only the aim is to disable communication related equipments and the computers.

Mobile and camouflaged targets like air defence radars, missile complexes, mobile troops, naval vessels over the high seas are good examples of the target system that can be attacked with E weapons. Location of such targets could be carried out through the use of Electronic Support Measures (ESM)) and Emitter Locating System (ELS). Once located, the slow moving targets could be attacked easily because they could not get away easily from the footprint of the bomb.

Army targets would be difficult to detect because those would be heavily camouflaged and do not radiate overtly. These could only be detected by tracking the Unintentional Emission (UE) also known as Van Eck radiation. Due to poor shielding electronic emissions leak out from various equipments used at the war front. Detection and demodulation of the same could give adequate target intelligence to attack them with electromagnetic weapons with impunity. There is no doubt that smart emitter locator could locate the emissions from computer networking cables, superheterodine receivers, etc. Deployment of UAVs over the suspected target areas could reap rich dividends.

Delivery of E Bombs

Like any other conventional warhead, the E bombs would also have some given mass determined by the density of the internal hardware. Typically an E bomb will comprise an electromagnetic device, an electrical energy converter and an energy storage device for sustaining the electromagnetic device charge after separation from the delivery platform. A radar altimeter fuse could provide fusing for air burst bombs. The fraction of the warhead could be 15 per cent and the rest of the space could be utilised for the electromagnetic device and other supporting elements. Like conventional warheads, E bomb could be fitted on a cruise missile, conventional missiles or on an aircraft.[14]

Surface-to-Surface Missile: E bomb placed on a surface-to-surface missile would comprise an electromagnetic device, an electrical converter and an on board storage device such as a battery. The battery is drained as the weapon is pumped. Missile on board fusing system would detonate the electromagnetic device.

Cruise Missile: Arming a cruise missile with E bomb may well impose certain restrictions. The size of the priming current and its battery may be the root cause of these limitations. In a cruise missile the bomb could be tied to the navigation system, in an anti-shipping missile the radar seeker and in air-to-air missile, the proximity fusing system. The warhead to launch-mass of the weapon should be 15 percent to 30 percent.

Delivery by Conventional Aircraft: Delivery of these types of weapons by aircraft could score over other systems. Because the launch aircraft having its own power system could prime the weapon optimally without any loss of power. In such a situation the bomb's capacitor bank can be charged by the launch aircraft en route to target, and after release, a much smaller on-board power supply could be used to maintain the charge in the priming source for weapon initiation. Every aircraft capable of delivering a standard weapon should be in aposition to deliver an E bomb. And should the weapon ballistic criteria remain the same, then no software changes for the delivering aircraft would be required. Stand off delivery of these kinds of weapons could be the only answer because of its lethal radius. Cruise missile or delivery by any other type of missiles would pose no problems. Delivery of E bombs from conventional aircraft would require careful planning lest the aircraft becomes victim of the bomb itself. Fire and forget type guidance system would be suitable for all types of air delivery weapons like air-to-air, air to surface and glide bombs. The launch aircraft must gain sufficient separation of several miles before the bomb detonates. This could be carried out either by toss bombing and delivering a glide bomb. The recently built GPS satellite navigation kits for conventional bombs and glide bombs has increased the accuracy of the weapon to a large extent. Therefore toss bombing both from low level and high level could be the answer. Importance of glide bombs could be manifold. The glide bomb can be released from outside the effective radius of target air defences, minimising the risk for attacking aircraft. Large stand off distance means that the aircraft can remain well clear of the bomb's effects.

The bomb's autopilot may be programmed to shape the terminal trajectory of the weapon, in a way that the target may be engaged from the most suitable altitude and aspect[15].

Search and Kill by UAV: An easy electrical kill could be achieved by using UAVs armed with emission locator and E weapons. Because of their inherent advantages an UAV could loiter over a target for a considerable period of time away from the base and once the emission is located distinctly, it could launch its on-board weapon.

Limitations of Electromagnetic Bomb

Like all other weapon systems, E weapons also suffer from their inherent limitations. Despite being an area weapon, the E weapons will demand better accuracy for their delivery because it would be important to illuminate the target area completely to achieve best results. Therefore though it is not difficult to calculate the area of its effectiveness yet it suffers from better CEP syndrome. Electromagnetic field strength achievable at a certain distance from the application would be the prime consideration.

Vacuum tube equipment based on Thermionic technology is extremely resilient to the electromagnetic weapons effect than solid-state technology. Therefore a weapon designed to destroy solid-state computers and receivers, may cause a little damage to a thermionic device. Early 1960 Soviet equipment falls in this category. This leads to another limitation of the E bombs which pertains to difficulties of kill assessment. The enemy may fool the attacker by switching off the emitters prior to an imminent attack and once the attack is over they may switch on the emitters for its usual functioning. Absence of emissions therefore would not indicate the success or failure of the attack. Nonradiating electronic equipment would also get affected due to an electronic attack but success of the same could not be assessed correctly also. Van Eck radiation detectors would be the only answer.

Atmosphere plays a leading role towards the ultimate effectiveness of the E weapons. Lethal coverage of an E bomb will depend on atmospheric propagation. The decays in lethal effect with increasing distance within the atmosphere will be greater due to higher frequencies and significant absorption peaks due to existing water vapour and oxygen. These will reduce the effect of HPM weapons to smaller radii than are ideally achievable in the K and L band[16].

Electrical collateral damage could be another limiting factor of this type of equipment. However no international treaty or legislation has been brought forward to curb its application.

Defence Against EM Bombs

In any future war our defence forces are going to face some kind of EM weapons and it would be necessary to pay attention towards the defence against this emerging threat of the future. No air defence system could provide total safety from air attacks and therefore those systems that could suffer probable damage by such a new menace must be electromagnetically hardened.

Faraday Caging could be one method of this hardening process. Cage the entire system in an electrically conductive atmosphere to avoid being exposed to electromagnetic radiation. This can prevent the electromagnetic field from gaining access to the protected equipment. But that would be next to impossible and certain amount of damage cannot be ruled out because the equipment would have to breathe and maintain contact with the outside world. This will give it away and its protective shield. To ensure better security the lines entering the equipment could be fed through an arrester.[17]

Limitation of Hardening

There is no doubt that system hardening is an absolute must to avoid damage to any part of the electromagnetic equipment. But hardening has its limitations. Conceptually the E weapon menace is at its infancy, as such no one really knows whether equipment can be hardened totally or some part will remain vulnerable to electromagnetic attack. And if one part gets affected, what would happen to the equipment as a whole? Considering this futuristic threat, the newer generation equipments could be hardened to an extent at certain cost. But it would be next to impossible to harden the older equipment. The damage is sustained mainly by the wounding semi-conductors and as such the equipment does not get destroyed completely and may keep on working intermittently, which would be more disturbing than a complete breakdown.

Electronic Combat Operations

Command of the electromagnetic spectrum should be achieved as quickly as possible to carry out subsequent operations with other conventional weapons. Futuristic wars would face this menace. Unless effective steps are being taken, a nation could become a hapless victim. Hard and soft kills would render the command control ineffective, the economic system could suffer most grievously, subject to the lethality of the weapon and the hardness of the target. An attack may not render the entire equipment totally useless but certain affected parts will make it partially serviceable which could be dangerous.[18]

The massed application of electromagnetic bombs in the opening phase of an electronic battle will allow much faster attainment of command of the electromagnetic spectrum, as it will inflict attrition upon electronic assets at a much faster rate than any conventional means.

Electronic combat operation using E bombs involves attacking radars, command control system, and air defence weapon systems. These need to be attacked first to paralyse the enemy decision-making ability or to ingress into its OODA loop.

The idea would be to achieve hard or soft electrical kill at the very opening round, followed by conventional attack with a fair amount of impunity. Presently the ARMs (Anti Radiation Missiles) are used for Suppression of Enemy Air Defence (SEAD) operation which are not only expensive, but have many other disadvantages, like single weapon for a single target, etc. But E bomb can be targeted against multiple targets, achieving much better results than the ARMs. That is why Calo Kopp in his article " The Electronic Bomb" has termed the E bombs as "Weapon of Electrical Mass Destruction (WEMD)". He has gone to the extent of explaining the horrendous effect of such weapons system not known to mankind so far[19].

However, as the bomb is not very well known, and only scrappy evidence of its possible use in Kosovo war has emerged it does not allow us to judge its actual capability. Exact calculation of force multiplication could not be carried out. But it is also true that an aircraft carrying a collection of E weapons capable of disabling a SAM site with collocated acquisition and fire control radar will have equal potential of many ARMs fired from multiple aircraft supported by the integral jammer force.

As such, it is seen that massed application of electromagnetic weapons for Electronic Combat operations will provide a much faster rate of attrition against hostile electronic assets, in comparison to any other conventional means. E weapons would allow much greater concentration of force.

Post-Gulf war strategic air campaigns are bound to follow the Warden model, which identifies five centers of gravity in a nation's war fighting capability. IW model can be formulated in similar fashion, that is to disable an opponent's fundamental information processing infrastructure as part of the Strategic IW campaign. But this is yet to happen anywhere so far. Warden model is explained as follows:

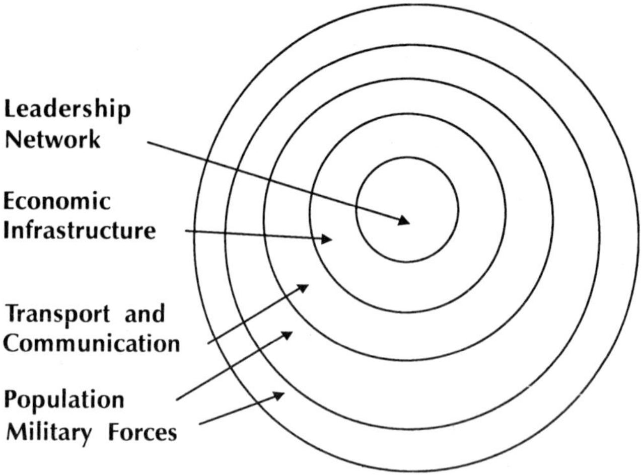

The innermost ring in Warden's model comprises the leadership network, which comprises government bureaucrats, and civilian and military C^3 systems. The effectiveness of attack on this inner ring will produce devastating results by disconnecting the decision-making authority from the rest of the war making machines. Of particular interest in this context of strategic air attack is that while E weapons are lethal to electronics they have little or nil effect on the human beings. Such peculiarity is not inherent to other weapon systems either conventional or nuclear. The cost of hardening existing computer networks is prohibitive, as is the cost of replacement with hardened equipment. Therefore, the use of E weapons against government facilities should be extremely successful[20].

Electronic Counter Air Operations

Airfields are geographically located areas housing innumerable electronic systems both on base and inside an aircraft and would be ideal targets for E weapons. Attacking airfields with E weapons could make them totally useless to mount both offensive and defensive operations. Attack on electromagnetic spectrum will disable communication, air traffic control facilities, navigational aids, air defence setup and other operational support equipment. The aircraft on ground will also become easy targets because the modern generation aircraft rely heavily on electronic equipment. It would be impossible to safeguard them considering the immensity of the problems about their hardening. Conventional blast hardening will not be effective. Electrical power cabling and communications cabling will carry electromagnetic induced transients into most buildings. Therefore Electronic Counter Air Attack could give the initial advantage of having a Favourable Air Situation.

Electronic Battlefield Air Interdiction

Win or lose situation in any future ground battle will depend upon the mobility and communication capability of any army. The armoured regiments are heavily packed with electronic and communication equipment supported by air defence weapon systems. Coordination and control of them will be the essence of the battle. An attack on the armoured concentration will render this force rather harmless. The effectiveness of their fire control and communication systems would be destroyed. The best tactics would be to attack the armoured formation with E bombs and create confusion. Follow up with an attack with conventional weapons.

Electronic Air Defence Operations

Interception is the buzzword in the air defence parlance. You use whatever is available in your arsenal—missiles, interceptors or guns—to deter the strike force to deliver its weapons in own territory. These platforms are used to counter enemy raids one by one. But it is beyond the comprehension of the air defence community that an area weapon like E weapon can be used against an incoming raid. The E weapons can be effectively used in all air defence operations provided the weapon can be made sufficiently compact to be loaded on an air-to-air missile. The enemy formation depending heavily on electronic on-board equipment could face a most novel challenge by losing their eyes and ears to carry out any successful attack and return safely to their bases. Their Nav attack radar could misread, mission computers could malfunction, digital engine control computers could fail, and communication equipment and electronic flight controls could become useless thus making the aircraft totally vulnerable to attack by defending fighters or conventional missiles[21].

Electronic Maritime Air Operations

Modern warships or a carrier group depend mainly on electromagnetic spectrum for their operations over the high seas. An E weapon could render that surface force totally ineffective if caught within its lethal radius. Thus making it vulnerable to any type of conventional attack.

Information Age Terrorism

India is constantly fighting terrorism and insurgency since independence and there is no doubt that our armed forces are the most experienced in the entire world in this respect. The Indian armed forces are yet to counter this novel and silent way to spread terrorism. To counter these 21st century insurgents, Indian defence forces should include information age terrorism in their doctrine and tactics before the troops in the field are surprised by these novel enemies.[22]

Indian Electronic Weapons

What is new about IW is the centrality given to these previously important, but secondary aspects of warfare. Also we cannot deny that IT in the civilian economy and national infrastructure has overshadowed the military in many a way. Disruption of the electromagnetic spectrum would not be tolerated by the Western powers at any cost. They are bound to bring international legislation to curb the desire of the aspiring nations to develop such capabilities to manufacture EMP weapons. These weapons do not cause any collateral damage and are not very well known to the world community. But it can cause havoc to many systems belonging to the nations riding high on the IT tigers. The Westerners call it 'collateral electrical damage'. The imagination of many a nation is yet to catch up with this awesome capability and understand its legal and other effects. India termed as hegemonistic among her neighbours and long harbouring the idea of force multipliers, could take a quantum jump by developing these weapons and perfecting their delivery system. Indian technology is quite mature enough to address this not too difficult a challenge. Force multipliers have been the dream of the military planners of India for many years based on AWACs and air-to-air refueling system. But procurement and system integration would take many years. IL76 (Moss) based AWAC is a far cry from the actual Indian need. Whereas, a force multiplier or a clean winner could be the EMP weapons[23]. No matter what Pakistan could procure or develop in future will not be in a position to match this capability. Air Defence system built around EMP weapons could nullify many a Chinese threat of the future. The situation, advantage India, cannot be lost. It is the right moment for India to develop, build and stockpile these cleanest of the bombs before the 'collateral electrical damage' becomes a rallying point for the highly vulnerable affluent nations and the next political battle is being fought by India at Geneva following the epic CTBT skirmish.

To begin with the Indian military must take the lead, harden their electronic equipment and provide thrust for development of EMP weapons. Infrastructure of DRDO could be pressed into action. The IITs and the Indian scientific community could be asked to participate. India does not lack technology, it only lacks scientific leadership. There is no doubt that the current technological base of India can produce EMP weapons and perfect their delivery system, before our neighbours wake up and cry wolf.

The revelation by Russia about the US testing EMP weapons in Yugoslavia could be a pointer towards another kind of war that mankind will experience in future. Helplessness against this type of attack can well be imagined only. The information hungry world depending upon real-time communication, internet revolution, electronics based entertainment industry, electronic sophistications of military equipment, over dependence on electromagnetic spectrum, would make the nations more and more vulnerable to EMP weapons. Hardening of equipments could create certain amount of safety but to build up all round credible defence against an EMP attack would be impossible. As the EMP weapons do not cause any damage to population or visible property of a nation the attack may go unnoticed by the world. Victim nations will understand and pay for any misadventure. That is why we would not mind calling these deadly clean bombs as the best force multipliers of the future[24].

Notes:

---

1 David A. Fulghum: *Microwave Weapons Await a Future War,* Aviation Week & Space Technology, Washington June 7, 1996. p30.

2 Ibid. p30.

3 Ibid

4 "Space Operations Orientation Course Handbook," Third Edition (Colorado 80909-6822, Air Force Space Command, Peterson AFB,,1 Aug 1993.)

5 Richard J Harknett. "Information Warfare and Deterrence". *Parameter,* Autumn 1996, ". ...destruction of societal connectivity need not be limited to specific attacks on personal electronic or institutional records. It may be conducted on a broader scale, where electronic connectivity is indiscriminately targeted. ......But little expertise is required to manufacture enough of a magnetic field in a directed fashion so as to put at risk the circuitry of an individual computer or computer system. Weapons such as HERF (high energy radio frequency) guns and EMP/T (electromagnetic pulse transformer) bombs are not only conceivable but may already have been built." p96.

6 'Space Operations Orientation Course Handbook", Third Edition, (Colorado 80909-6822 Air Force Space Command, Peterson AFB, 1 Aug 1993.)

7 Timothy L Thomas:. "Deterring Information Warfare: A New Strategic Challenge" *Parameter,* Winter *1996-97,* pp81-91.

8 NHQ Publication *Information Warfare* Vol 6, 2000.

9 Carlo Kopp "The Electromagnetic Bomb – a Weapon of Electrical Mass. Destruction" < http://www.cs.monash.edu.au/-carlo/ > pp 10-12. While

discussing vircator, Carlo Copp opined that though the technology is not mature as yet, but it would not take long for the scientific community to give it a real shape. He also argued that this would be the most potent weapon of the future.

10 **Ibid.**

11 Ibid p14.

12 NHQ Publication *Information Warfare* Vol 6, 2000.

13 Kopp n9.

14 John Tirman.. "Collision of a particle from a Neutral Particle Beam Weapon with an air molecule". *The Fallacy of Star Wars By Union of Concerned Scientists*, New York: Vintage Books, 1983 Chapter 5, p11.

15 Joseph Goldblat and David Cox. *Nuclear Weapon Tests Prohibition or Limitation*" Sipri, Oxford University Press 1988. pp 240-241.

16 Robert Jastrow *How to Make Nuclear Weapons Obsolete*, London: Sidgwick and Jackson 1985. pp89-99.

17 Kopp n9.

18 Peter D Feaver "Blowback:Information Warfare and the Dynamics of Coercion." *Military Review,* Sep-Nov 1998, pp26-30.

19 Carlo Kopp "The Electromagnetic Bomb – a Weapon of Electrical Mass Destruction" < http://www.cs.monash.edu.au/-carlo/

20 Laos Nicholas K "Information Warfare and Low Intensity operations." *Perceptions,* Jul-Aug 1999, pp174-195.

21 Clemmons Byard "Cyber Warfare: Ways, Warriors and Weapons of Mass Destruction" *Military Review*, Sep-Oct 1999, pp35-45.

22 Arquilla, John et al "Information Age Terrorism" *Current History*, April 2000, pp179-185.

23 Ellis R Evan "Organisational Learning Dominance: The Emerging Key to Success". *Comparative Strategy* Apr-Jun 1999, pp 191-202.

24 Roger C Molander, Andrew S. Riddle, and Peter A Wilson. "Strategic Information Warfare: A New Face of Warfare" *Parameter*, Autumn 1996, pp81-92.

7

Directed Energy Weapons

Considering the capabilities of laser, the United Nations decided to prevent its use against human beings and a protocol[1] was signed to limit the use of laser blinding weapons. As it stands, NATO nations are bound by this protocol and hopefully will restrict their use of laser weapons honouring the code of conduct dictated by the United Nations. These nations are to use laser against, electro-optical sensors. But the development process to this effect belies all hopes of the United Nations. Laser weapons have already been dubbed as inhumane. Yet, the developments of these weapons are progressing unabated. Certain defence analysts have already opined that the concept of nuclear weapon is as outdated as the bows and arrows. Directed Energy Weapons (DEWs) would replace the nuclear monsters soon enough and take the center stage, probably proving the legendary HG Wells to be correct. In his book *The War of the Worlds* Wells talked about the invading Martians with heat ray guns. These ray guns of Wells are chillingly similar to the real laser weapons of today. At that time on, DEWs may have been the prerogative of comics and movies. But the superpowers of the pre-cold war era took it up on themselves to develop these super killers. Clandestine research developments went ahead under strict veil of secrecy. Reports that are appearing in various magazines bring out the general facts to a large extent about the state of these weapons and the state of their development and the goal. This does not indicate in any way that the advanced nations are ready to subscribe to the United Nations protocol regarding DEWs and their future use.

Concept of the DEWs may have been as early as in 212 BC in Greece, when Archimedes used polished mirrors to direct sunlight on the sails of Roman ships while defending the city of Syracuse. And it was not until 1890 that HG Wells resurrected this concept in his book, *The War of the Worlds*. And taking a leaf out of the history, Ronald Reagan, the US President, on 23 May, 1983, urged the development of a systems that could intercept and destroy strategic ballistic missiles before these could reach the US soil[2] or that of the US allies. He directly copied the idea from the star wars. Enough funding was promised despite loud protests from various quarters. Subsequently a report appeared in the IDR August 2000 issue that the DEWs are well on their way towards actual deployment[3].

High Energy Lasers

Laser stands for Light Amplification by Stimulated Emission of Radiation, is nothing but an atom or a molecule stimulated to reach a very high-energy state. This is done by various means like chemical reaction, electrical currents or intense light. When an atom falls to a lower energy state, it emits a photon. This photon has some finite energy and it is capable of colliding with another atom in a high-energy state and extracting another identical photon from it. This is called stimulated emission[4]. Stimulated emission of lights occurs when electrons are forced by incident radiation to add more photons to an incident beam. In a laser there are mirrors, which reflect these photons back and forth through the medium, the emission gets amplified and the escaping photons are known as the laser beam. This output beam is a light of continuous wavelength, which depends on the lasing medium being used. Laser light is highly monochromatic because it is generated by electron transitions between two narrow energy levels. As a consequence, laser light can be focused to a spot, less than 1 mm in diameter[5]. Another outstanding feature of laser light is its strong collimation i.e. the parallel emergence of light from a laser window[6]. Therefore, it may be clearly understood that a laser weapon works by depositing intense heat energy on the target and destroying it by burning or causing fracture. Laser can be used for two types of attack, thermal and impulse. In a thermal kill, the target is destroyed by a long dwell time of the laser and burning through it. On the other hand, if a very high power laser pulse is used, it vapourises a small layer of the target skin and its surrounding air. These super heated gases expand explosively sending shock wave in to the target. Such a shock wave can destroy the skin, and is known as impulse kill. A particle beam could bore a hole in the outer shell of a missile, a satellite or an aircraft and cause serious damage to electronics inside or explode the high explosives trigger of a nuclear weapon. Such damage requires about 10^3 joules/cm^3 deposited on the target[7]. The total energy of beam that has an angular divergence of a micro radian and can deliver that amount of energy can be generated by an accelerator that produces a particle beam pulse of 1000 amperes flux accelerated to 1 GeV (Gega electron volt) that lasts for 100 msec[8]. (One electron volt may be explained as the energy required to bring an electron from 0 volt to 1 volt.)

Exoatmospheric uses of electron and proton beams will have their problems[9].

- Spreading of the beam due to the coulomb repulsion of similar charges. An electron beam of 1000 ampere, 1 GeV with an initial radius of 1 cm, will be 15 meters in diameter at a distance of 1000 km away[10]. A proton beam can spread upto 18 kms in diameter at a similar distance. The target will remain unharmed unless the beam can be concentrated on it. Therefore beam concentration is a prerequisite for this type of weapons.

- Charged particle beam bends due to the geomagnetic forces. A beam like the one described above will have a radius of curvature of about 100 km and therefore will never reach the target. It is rather impossible to calculate the exact position of a charged particle beam at large distance, because of the uncertainty of the strength of the geomagnetic field in the intervening space between the weapon and the target[11].

- During their generation Neutron beams also disperse very badly, generation because of the properties of nuclear matter. As a matter of fact a neutral hydrogen beam can be produced but will have dispersion of about 10 urad and will require energy like 10^{10} joules. It was considered difficult to produce such level of energy but presently things are changing.

Limitations of Laser Beam

Inside atmospheric conditions the beam suffers from absorption and scattering, and becomes defocused and bent because of the properties of air. Whereas, in the vacuum of outer space the beam suffers only diffraction, which is inversely proportional to the size of the laser. There is an upper boundary of power density (10^7 /cm^2) that a laser beam can transmit inside atmosphere[12], because laser light of very high intensity can cause electrical breakdown of the air, which in turn disrupts the beam. Moreover, laser weapons efficiency while operating in earth's atmosphere depends on local weather conditions and the presence of dust, smoke or aerosols. Application of laser for target detection and tracking would have to take these factors into consideration.

A scientific calculation based on facts can prove beyond doubt that a laser beam may not be able to melt or vapourise a metallic target very easily without disturbing the air, because the power density levels required, tend to exceed the breakdown threshold of air. But there is no doubt that the most effective way to damage a target with a thin metallic skin like aircraft or missile will be by impulsive loading because these structures are already under severe stress. Any addition to this stress need be supplied to exceed the acceptable stress of the material. Thermo-mechanical damage can be caused to a target if it is exposed to a repetitively pulsed laser beam; each pulse will deposit plastic energy in the target, until an amount equivalent to the fracture energy is reached. This repetitive pulse beam exploits both thermal weakening and mechanical damage and will melt through.

Despite the UN protocol it can be safely estimated that high-energy lasers will be used in the near future for attacking targets like aircraft, satellites, ballistic missiles, and cruise missiles. The design of these new weapons systems will depend on the vulnerabilities presented by the target. To cause maximum damage, the laser wavelength, pulse shape, energy, and beam diameter should be chosen correctly. The thermal damage to a target need not necessarily be melting. It can cause thermal damage mostly due to overheating. The energies absorbed by the target of $\sim 10^3$ joules/cm^3 per pulse delivered in pulses of several tens of microseconds over a small area of the target will cause impulsive failure of a thin metal target[13].

There are four major types of lasers for weapon system applications. These would have differing wavelengths and differ in their output power. All of them could be employed in potential weapons systems. The types of laser are:

- Gas Dynamics Laser[14]
- Chemical Laser[15],
- Solid State Laser[16]
  - o Ruby[17]
  - o YAG[18]
  - o Glass Laser[19]
  - o Semi Conductor Laser[20]
  - o Heat Capacity Laser[21]
- Free Electron Laser[22]

Despite production of high-energy laser, the fundamental requirement would be strong coupling action of the laser light to the target in order to cause any damage. This would depend on the factors like wavelength, target material, and the aspect angle of the target to the beam. That is why a military application will have to take into consideration reflectivity of the beam for beam control and best absorption of the light by the target to cause maximum damage. IDR reported that:

> " US Department of Defense currently funds three kinds of device technology for High Energy Laser weapons; chemical, solid state and free electron. Chemical lasers use a chemical reaction (combustion) to provide the pumping energy. Weapon class chemical lasers include types using hydrogen fluoride(HF) and deuterium fluoride (DF) and chemical oxygen iodine lasers (COIL). These have achieved power levels of several megawatts with good beam quality." *IDR* August 2000.[23]

Particle Beam Weapon

History of particle beam dates back to 1950's, even before the laser made its theoretical appearance. Scientists were able to accelerate charged sub atomic particles to very high speed in an accelerator[24]. And all such devices generate their destructive power by accelerating sufficient quantities of these subatomic particles or atoms to velocities near the speed of light and focus the energy into a high energy beam. Each particle moving in the beam would have its kinetic energy due to its own mass and motion. Presently particles used to form this beam are electrons, protons and hydrogen atoms. In the case of hydrogen a single electron and a proton combination form a neutrally charged atom. Because of being neutrally charged, hydrogen atoms would not be susceptible to bending by the earth's magnetic field, which is a problem of weaponisation of particles beam (PB). In the atmosphere charged particle beam neutralises itself by colliding with air molecules, effectively creating enough ions of the opposite charge to neutralise the beam.

To form a particle beam, an electronic accelerator should have a diode in a vacuum. A very high electrical voltage is applied across the diode, the negatively charged electrons that are emitted instead of being collected at the anode are made to pass into an accelerator. This accelerator has suitable arrangement of electrical and magnetic fields, which impart high acceleration to these electrons and can focus them also. This is a simple linear accelerator[25]. Similar process can be achieved in a circular accelerator with the electrons accelerating in circles of increasing diameters, which is known as cyclotron. Energy of the beam is expressed in electron volts, the beam current in amperes, and power of the beam in watts. The power of a PB is the rate at which it transports energy. That would give a clear idea about the rate at which the energy will be deposited on the target. The electron and protons of the beam are deposited on the target, which collide with the atoms of the target, and break the atoms like snooker balls. The area of the target where the beam has been focused will get heated up rapidly and the target will explode.

There are two types of Particle Beam Weapons (PBWs)—charged particle beam weapon and the neutral beam weapon. The charged particle weapons are for use within atmosphere (endoatmospheric) and the neutral particle beam weapons are used in space (exoatmepheric). Both these weapon systems will have different characteristics of their own[26].

The fire control, procurement and tracking system of PBW must acquire and track the target, point the weapon at the target, fire the beam at the appropriate time and assess target damage. The miss distance has to be calculated and correction has to be carried out on time. Advantages of PBW[27]:

- **Beam Velocity**: Being the speed of light (186,000 miles/sec), computing and the aim point for a moving target like an incoming missile become simpler. It would be difficult for the target to carry out any evasive maneuver.
- **Beam Dwell Time:** In an endoatmospheric weapon the target can be destroyed instantly provided the beam could be held even for a short period of time.
- **Rapid Aim Capability:** Because of its high velocity the beam can be aimed on a number of targets at very short range of time, by means of a magnetic field. Varying the current would change the magnetic field intensity which could deflect the particle beam.

- o **Beam Penetration**: Sub-atomic particles can penetrate a target. It is not restricted to surface effects as in the case of laser. It could damage the internal components of the target and even cause an explosion by implanting massive quantity of energy. It would be rather impossible to defend the target against such an attack. No amount of target hardening will have any effect.
- o **All Weather Capability**: PBW has an added advantage over the laser. It has an all weather capability. Laser beam can get degraded if there is fog, cloud or rain but it does not affect the PBW. In the atmosphere as the particle beam travels, it produces an ionised channel, which tends to keep the beam together and does not allow it to disperse.

Power Supply and Accelerator Technology

This is no doubt a difficult problem for the PBWs. To operate an endoatmospheric weapon, tremendous amount of electrical energy is required over a very short period of time. As power is energy divided by time, large amounts of energy over a short period of time translate into extremely high power level. Lightweight components capable of producing million of watts will be the actual requirement of the weapon system. In addition, the accelerators create the high-energy particle beam. It is composed of a source of ions (electron, protons or charged atoms) a device injecting the particles into the accelerating section and the accelerating section itself. There are various types of accelerators, like linear, circular and radio frequency linear. As reported, Advanced Test Accelerator is designed to generate 50 Me V (million electron volt) beam with 10,000 amperes of current in pulses of 50 nano second (50 billionths of a second)[28].

High Power Microwave Weapons

High Power Microwave weapons were on the drawing board for a long time. But for the first time in Kosovo war some testing could have been carried out, as claimed by the Russians. It seems that the weapon can be delivered by cruise missile or by any other stand off platform and can be exploded within yards of a target.. The explosion would be followed by generation of high power microwaves that can disable electronic circuitry in computers and communication systems. The electromagnetic pulses could create havoc to anything that is electronically operated. The EMP, which was noticed during the high altitude nuclear explosion, has at last being perfected by a team of US scientists probably at Los Alamos and may add a new and unknown dimension to a future war[29]. The technology that would be used to generate the EMP is:

- Explosively pumped flux compression generator
- HPM devices based on Virtual cathode ray oscillator or vircator
- Propellant driven magneto hydrodynamic generators

Like all other DEWs, defence against such weapons system would be very difficult to install because the capabilities of such a menace is not yet known exactly. It is very well to claim that certain amount of electronic hardening would be the solution but there are some serious limitations. It is not known whether equipment can be hardened totally or some part will remain vulnerable to electromagnetic attack. And if one part gets affected, what would happen to the equipment as a whole. Also it is quite clear that this would be the route to gain command of the electromagnetic spectrum so as to carry out subsequent operation with impunity using other conventional weapons. Subject to the lethality of the weapon, command and control system would become ineffective and cause havoc to economic system. Effect of a massed attack with such weapons at the opening phase of the battle can only be imagined.

DEWs of USA

"Lt Gen John Costello commanding general of the US Army's Space and Missile defence command describes DEWs as a potentially revolutionary addition to the battlefield of the future."[30] During the frightening period of the cold war arms race, both the super powers generated a lot of heat by claiming the development of death ray and other DEWs. But since then it had been the prerogative of only the USA to spend enough amount of money and time to gain superiority by gaining access to these frightful and seemingly innocent weapons systems and call for a global nuclear disarmament. It was way back in Falkland war of 1982 that the UK, an US ally, employed a laser dazzler and now it has come to notice that Chinese People's Liberation Army have deployed weapons like ZM87 laser interference device that could cause short term blindness at ranges upto 10 km. PLA's Type 98/WZ-123 main battle tank carries a turret mounted installation of a laser equipment which could be used against both ground based and air borne optical and electro optical sensors[31]. It is understood that the US development of the DEW is progressing unabated despite many obstacles. And there is no doubt that the US will forge ahead in this respect considering the pace of the development brought out in the IDR. High Energy Laser Executive Review Panel of the USA, headed by the Deputy Under Secretary of Defense for Science and Technology, clarified in their report that[32]:

- High-energy laser systems are ready to meet some of today's most challenging weapon application.
- And this laser weapons will offer the US the potential to maintain a technological edge over adversaries for the foreseeable future.

This is more or less an accurate statement because despite having a solid base in the field of physics, breakup of the Soviet Union has put many impediments for the growth of this field in present day Russia. Main problem is the funding for fundamental research. The US committee while commenting on the laser weapon development identified few areas that require immediate attention and to be supported by adequate funding[33]:

- Solid State Lasers of US Army: In this field the aim would be to develop and demonstrate coherently phased fiber arrays; scale up heat capacity lasers to powers of 100KW and beyond.
- Chemical lasers (USAF) Tactical weapons based on Chemical oxygen iodine laser (COIL) together with hydrogen fluoride (HF) and deuterium fluoride. Evaluate the proposed Air Borne Tactical Laser.
- Free Electron laser.
- For beam control develop illuminators, sensors, mirrors, etc.
- Determine lethality and effectivity of the pulsed versus continuous wave lasers.

Out of the many, USAF research for the Airborne Laser (ABL) [34] and Space Based Laser (SBL) System have advanced on the determined path. Large Aircraft Infrared Countermeasures programme (LAIRCOM) is aimed at arming C-17 and C130 air lifters, together with KC 135 tankers. USAF would be ready to deploy at least seven Boeing 747-400Fs with capabilities to engage ballistic missiles during their boost phase from a stand off distance of about 400 km[35]. The system is expected to be functional during the financial year 2005 and will hopefully be completed by the year 2009. The Space Based Laser (SBL) deployment will take longer than ABL. The entire system would comprise 20-40 satellites at altitudes of 700-1300 km for global coverage. The system would be able to engage any ballistic missile with a range of more than 120 km during its boost phase[36]. Each satellite would be equipped to carry acquisition, tracking and pointing system using low power illuminator and a high energy laser with a range of more than 3,000 km would engage the missile. The system fits all the QRs of the much talked National Missile Defence (NMD) of USA.

US Army's joint programme with Israeli scientific community which began in 1996 has brought about the development of the Tactical High Energy Laser (THEL), which could destroy a Katyusha rocket in flight carrying a live warhead[37]. On its success Lt Gen Costello commented that " we have just turned science fiction into reality". But it seems that the statement cannot be taken just on its face value. There is subtle warning about the US emergence as the sole leader in the field of DEW without a challenger to be seen anywhere.

DEWs are not mentioned in the Indian Defence environment because it is more or less taken for granted that such weapon systems are beyond our means and that is why they do not feature in any of the military studies. Indian military may not be thinking of application of laser beyond laser ranging and laser designation. Recently a study has been ordered by the COSC to look into the feasibility of the DEWs. DRDO, the only research organisation in India serving Indian defence purposes, has certain research projects on laser weapons but it is not known what is the state of that research and whether any weapon system is developed for deployment with the armed forces. But before the Indian armed forces can lay their hands on the DEWs and employ them, they may undertake defensive measures.

Defensive Tactics Against Directed Energy Weapons

As has been already described DEWs include lasers, Particle Beam (PB) and High Power Micro (HPM) wave weapons. These weapons produce casualties and damage equipment by depositing energy on the target. Whereas, conventional weapons rely upon kinetic/chemical energy of a projectile to destroy the target, DEWs depend upon subatomic particles or electro-magnetic waves impacting on the target at the speed of light and are expected to cause damage to soft targets like personnel, soft components of hard targets, like optics. It is expected that the equipment fielded in future wars will be built with certain amount of hardening against DEW attacks and older equipment may be refitted with protective devices. But presently the armed forces personnel should be taught to defend them against these weapons.

Laser systems will definitely be fielded by the enemies of Indian armed forces in some form or the other. There are increasing evidences of the presence of laser devices in the inventories of all the armed forces, like target designator or a laser range finder. The most probable target of laser weapon will be optical and electro-optical systems, and the personnel behind the sights. A laser beam entering a direct view optical system such as a telescope will have its power magnified. Anyone looking through the system will suffer burns to the eyes. The severity of the burns, the permanence of the damage and the time required for healing will depend upon weather conditions, the intensity of the laser, magnification of the optical laser and the duration of the eye's exposure to the laser. Eye injury may range from temporary flash blinding and mild burns to total blindness. A soldier subjected to this type of injury may be incapacitated and unable to aim his weapon. It is anticipated that a laser weapon will fire at a target for a split second and aim at another. Laser weapons fired against individuals could cause burn. As the eye is more sensitive during night, laser energy entering the eye during darkness will have a greater effect than it would during daylight. Certain form of laser will be hazardous to soldier's eyes even though the laser cannot be seen.

A laser beam entering an electro-optical device, such as night vision sights or thermal imagery device, will deposit its energy in the form of heat on the sensor screens inside. There are chances that this can burn out the screen, making the device useless. Any uncovered glass surface (eyeglasses, vision blocks, or binoculars) has the potential to attract or alert an anti-electro-optical weapon's target acquisition system. However, it is not precisely known about the effect of a laser beam on human beings.

Avoiding Anti-electro-optical Weapon Systems

- Fire artillery, mortars or direct fire weapons to suppress known or suspected anti-electro-optical weapons locations. Smoke rounds are very good for temporarily defeating laser devices.

- While operating from static positions within line of sight of known or suspected enemy locations, minimize the exposure of glass surfaces in the direction of the enemy by positioning vehicles and weapons in covered or concealed positions.
- When there is a possibility of the exposure of many glass surfaces, block the line of sight between friendly forces and known or suspected enemy locations with smoke or by planning routes to minimize this exposure time.
- Sound tactics will prevent enemy from locating weapons and prevent subsequent attack by laser devices.
- All devices that have external glass surfaces and are not in use should be covered or shielded until needed. Even vision blocks and headlights can alert anti-electro-optical weapon tagged acquisition systems, so these must be include when taking protective measures.
- Reduced number of personnel in observation, will reduce possibility of injuries. Use of night vision goggles and thermal nights sights when possible to protect these observers.
- Tubular extensions over optical lenses will lessen their chances of being detected except from almost head on.
- Low energy anti-electro-optical weapons will work only if they have line of sight to their target. However, smoke, fog, snow and dust will degrade their effectiveness. Another countermeasure would be to cover part of your optical lenses with tape or some other type of cover. Some degradation of the viewing will occur but it would be a direct benefit by reducing your vulnerability.

Defence Against EMP Weapons: Electromagnetic Pulse can severely damage or destroy sensitive electronic equipment such as microchips, coils and fuses by overloading them with electrical current[38]. Any equipment containing electronic components is subject to damage or destruction from EMP attack. The amount of damage to equipment depends on its distance from the source of the pulse and type of hardening it had. An EMP attack lasts only for a split second and affects a large area. Protecting equipment against such an attack is extremely difficult. The only way to protect equipment would be to totally encase the equipment in a metallic shielding. Burying or covering with sandbags or other nonmetallic materials will not provide any protection. Terrain masking is also ineffective because EMP follows the curvature of the earth. While operating from combat vehicles, sensitive equipment not needed for use at the moment should be disconnected and moved to the center of the vehicle. Hatch cover should be kept closed. By doing this, only a minimum of equipment is susceptible to destruction and the remainder is available for use after the attack. Direct or indirect fire weapons within range should attack known EMP generating ground-based weapons.

Defence Against PBW: Particle beam is a directed flow of atomic or subatomic particles. When concentrated into a beam that can interact with a target can melt or fracture target material. But it is not known whether a particle beam weapon is ready for deployment by any country. The defence against the same will be similar to that of the other DEWs as discussed earlier. Commanders at all levels will have to bring awareness to their subordinates to this new dimension of threat that is emerging and will become reality in the next round of combat.

Impact of Directed Energy Weapons

RMA is gaining ground in the thought process of the military strategists of the entire world and there is no doubt that the changes are technology and information driven. While discussing these, Kapil Kak of IDSA mentioned that:

> " ...Technology change may well revolutionise warfare in the 21st century. Countries that can exploit emerging technologies and synergies the same with innovative operational doctrines and organizational adaptation could doubtless achieve far higher levels of relative military effectiveness.[39]"

Technology changes are round the corner and these would be more dangerous and more lethal than anything seen or experienced before. The directed energy weapons neither make any noise nor they inform about their deployment. The super observation capability of a nation may discover the design of an adversary but it would be very difficult to determine the time and direction of the attack and the effectiveness. For an example, the EMP weapons could write off the entire computer systems and throw the command and control in complete disarray. Few military analysts know, how to combat this new dimension of warfare. A complete paradigm shift is expected when the Laser and Particle Beam weapons are brought in. Nuclear bombs and the perceived threat from them vis a vis these new developments have to be analysed by the military thinkers and take effective action before being completely surprised. It would not be very easy to assess the impact of these weapons because of their peculiarities and the absence of collateral damage scenario. Collateral electrical damage is the surest possibility. It may paralise a nation for a few days but as long it does not affect a human being no one will raise a finger at the aggressor and make an issue out of it. The aggressor will achieve his purpose. Nuclear doctrines, mutually assured destruction, limited response, CTBT, MCTR etc., will lose their relevance in the backdrop of the silent revolution of DEWs. The Airborne laser system to be deployed by the USA, with capability to destroy a missile with any range beyond 120 km at boost phase, is not very far away from now. But the space based laser or particle beam weapon capable of punching a hole to any missile at a range of 3000 km or more can only be analysed to an extent.

Indian military planners must take these new threats into their force structuring or else they will face a new kind of battlefield unknown to them. The concept of DEW application is like the morning star on the horizon. It looks good but difficult to reach. The DRDO is in the process of developing some of these weapon systems but till such time these are made available to the military, the systems cannot be integrated with other weapons in the inventory of Indian armed forces. DRDO may carry on with their own research and will be welcomed by the armed forces if they produce something worthwhile and hand it over to the Indian military. But it would be consistent to suggest that Indian armed forces must have their own laboratories manned by dedicated scientists, to develop the futuristic weapons under proven military scientific leadership or else the technological world will drive past the door of Indian military and the RMA will be the word of the analysts only. It should be understood by the Indian military system that technology driven RMAs are brought about by weapons or systems exploiting powerful combinations of technologies, combined with supporting doctrine and organisation, input process being the technology breakthrough and the output being the RMA.

That the current technological breakthrough in relation to DEWs is real term RMA or not, cannot be defined readily now. Because only after a detailed strategic analysis, it can be assessed whether the new weapons can render obsolete or irrelevant one or more core component of a dominant player in the new dimension of the warfare. But it cannot be discounted either. The cyberspace-based technologies supported by the DEWs in a future conflict might inflict unacceptable damage sufficient to significantly alter the course of the conflict. Revolution in military affairs is however unclear and definitely not taking a predictive course. That is why a hypothesis can only be drawn from the present about the future.

History of Laser and Particle Beam Weapons

1958	Laser theory by Charles H Towns.
1960	First Laser and simple laser operations.
1960's	Carbon dioxide laser Invisible beam at wavelength of 10 micrometers in the infrared. C Kumar N Patel at Bell lab 8800 watts 750 feet beam.
1967	Gas Dynamic laser–scalable to high energy.

1969	Chemical laser reach high power.
1970's	Laser range finders and designators.
1970's	First military operational laser.
1973	USAF tested shooting down of drones with laser. The project was kept secret till 1982.
1975	US Army used 30 to 40 kW laser in a tank to shoot down helicopter drones at Red Stone Arsenal, Alabama.
1978	US Navy demonstrated moderate power chemical laser 400 kW to shoot down a TOW antitank missile.
1991	Alpha programme confirmed that a high-energy chemical laser could generate sufficient power to destroy a missile at a good distance. During that period it was demonstrated that large optics were possible.
1968-2000	USAF spent approximately $5.2 billion between fiscal year 1968–2000 on applications of directed energy weapons for offensive and defensive applications and plans to invest a further $4.6 billion by financial year 2007.
2000	USAF is continuing its development of the following projects[25]:

- o Airborne Laser (ABL) to be deployed on C-17, C-230 and KC-130. It would be known as Large Aircraft Infrared Countermeasure (LAIRCM). Also, the USAF will deploy seven Boeing 747-400 Fs that could engage ballistic missile at its boost phase from stand off ranges of more than 400 kms. The ABL passed its critical design review in April 2000.
- o Space based laser.This would involve a constellation of 20-40 satellites at altitudes of 700-1,300 km, providing continuous overlap of global coverage in all kinds of weather. The system will be deployed to engage any

ballistic missile with a range of 120 km or more during its boost phase. It does not require prior knowledge of the launch site. Technical details have been demonstrated.

History of PBW

1958	PBW development programme initiated by DARPA projects Seesaw.
1974-78	Chair Heritage Particle Beams for US Navy. Micro explosion fuel pallets–miniature versions of hydrogen bombs. Internal Confinement Fusion simulates effects of nuclear explosion. Accelerator technology and pulsed power systems.
1977	Soviet Union Particle Beam weapons for ballistic missile defense.
1983	Ronald Regan for the first time talked about rendering the nuclear missile obsolete by deployment of PBWs.

Requirements of Directed Energy Weapons

Operational Requirements for Directed Energy Weapons:

- Detection of target.
- Identification against countermeasures.
- Fire the beam or particular pulse of requisite shape at the target and maintain the same on it.
- It would be necessary to determine whether the target was actually hit or not. And carry out accurate damage assessment.
- Determine miss distance if any. Then correct for the same at given point of time.

DEW system must have the following :

- A laser platform like aircraft or satellite.

- Fuel and coolant stores.
- Facility to produce necessary electrical power.
- A power staging and switching system that may be necessary for the operation of certain types of lasers or accelerators.
- A system to control the tracker to the laser mirror or the beam bending magnets.
- Control and computer integration.

Some special requirements of the high-energy laser and particle beam weapons :

- An accelerator to produce a beam of particles of adequate energy and intensity that could damage the target.
- A pointing system for the laser mirror.
- A laser mirror or mirrors to focus and direct the laser beam.
- One or more laser cavities of appropriate power.
- Mechanism for beam neutralization before it leaves the weapon.
- Control system to steer the beam and aim it on the target. The entire system efficiency will depend on this particular requirement.

Notes:

1 Review Conference of States Parties Conventional Weapons Convention Vienna, 25 Sep-13 October, 1995," Adopt Ban on Use, Transfer of Blinding Laser Weapons"(Press Release United Nations DC/2525, 17 October, 1995).

2 Lisbeth Gronland and David Wright, *"The Star Wars Legacy"* Union of Concerned Scientists Issue Brief Vol 2 No 9 March 25, 1998.

3 *Jane's International Defense Review* 8/2000. p34.

4 RE Hunamel, *Electronic Properties of Materials*, New Delhi: Narosa Publishing, 1993; pp239-243.

5 Ibid p240.

6 Ibid p241.

7 William Guttridge and Trevor Taylor, *Dangers of New Weapon System*, (London and Basingstoke: The Macmillan Press Ltd 1983) *Kosta Tsipis, Directed Energy Weapons,* pp25-27.

8 Ibid p25.

9 Ibid pp25-29.

10 Ibid

11 Ibid

12 Ibid p26.

13 Ibid p27.

14 Lasers that employ electrically excited gasses as the active medium have operated at the highest efficiency. The gas discharges remain optically homogeneous and virtually there is no loss at the lasing wavelength even when the optical path length is several meters long. In this context, Carbon Dioxide Lasers[14] may be best examples. The output of this laser could be to the tune of 9 kw of power. An electron beam can only produce this type of energy. Monte Ross *"Laser Applications"* (New York: Academic press 1971)pp203-214.

15 Chemical Laser. Chemical oxygen iodine lasers together with hydrogen fluoride and deuterium fluoride are under development by the USAF for tactical deployment[15]. Ibid

16 Solid-State Laser. The solid state optically pumped laser produces the highest energy peak power. This is possible because of very high density of laser ions available in solid materials. Concentration of $\sim 10^{20}$ laser ions/cm^3 are typical for solid hosts. Whereas gaseous lasers usually operate with lesser than $\sim 10^{18}$ laser particles/cm^3. Moreover, solid laser materials can be excited very rapidly with high energy flash lamps to extract the stored laser energy in a great pulse of short duration. Ibid pp212-214.

17 Ruby Laser. Ruby was the first material in which laser action was demonstrated and has continued to be an important source of high power laser emission. It is an extremely hard and durable material with high thermal conductivity, permitting effective cooling of the laser rod. Laser actions are quite dependent upon the rod temperature, so an efficient cooling system is required for its continuous operations. Ibid pp215-216.

18 YAG. A short name of Yttrium Aluminum Garnet (YAG) doped with Nd^{3+} (Neodium) or material like chromium, erubrium, holium, cerium and titanium posseses many favourable qualities to make it a good source material for laser. It is hard, durable and has good thermal conductivity. YAG also can be produced with excellent optical quality. The cost and its limiting size may restrict the use of YAG lasers commercially but it may be used for weapon application. It may be pointed out that because of its unique qualities, YAG is the only solid-state laser material that has generated significant amount of continuous laser power, over 1100W at 1.06 mm. (wavelength). Ibid p218.

19 Glass Laser. Glass lasers can produce very high-energy pulses in short length of time. Lasing strength more than 5000 J is not uncommon. Barium crown glass and high silicate glasses are commonly used doped with Nd^{3} (Neodium). Glass laser rods over 72 inches in length have been fabricated with good optical qualities[19]. Ibid pp216-217.

20 Semi Conductor Lasers. Semi conductor lasers developed recently are going to replace the gas based ion laser and other solid state lasers. Power as high as 1000W can be achieved easily. Wavelength of these lasers can be from infrared to ultra violet range. Peltier or conventional cooling can be used for semi-conductor based lasers. Power consumption is much lower than the other types of lasers and this will help much-needed miniaturisation for weapons systems of the future. RE Hunamel, *Electronioc Properties of Materials*, (New Delhi, Narosa Publishing, 1993) pp245-252.

21 Solid-state lasers also include heat capacity lasers, fiber optic lasers and continuously cooled designs. Typical criteria of certain solid-state lasers are their tenability over a large bandwidth.. Titanium sapphire lasers have the bandwidth of 0.66-1.18 microns., whereas, alexandrite (a chromium doped crysoberyl, crystal can have the wavelengths from 0.7 to 0.82 microns. Ibid.

22 Free Electron Lasers are high efficiency lasers and can be tuned over a broad part of the spectrum. These lasers produce their energy by passing an electron beam through a structured magnetic field. Electrons change direction sharply because of alternate magnets placed in the array which have reversed polarity and that causes the emission of the energy. Ibid.

23 Marie Hewish, "Beam Weapons Revolution" *Janes International Defense Review,* August/2000

24 Emanuele Rimini, *"Ion Implantation: Basic to Device Fabrication,"* (Boston/Dordrecht/London: Kluwer Academic Publishers 1995,) pp33-39.

25 Ibid.

26 William Guttridge and Trevor Taylor, *Dangers of New Weapon System,* London and Basingstoke: The Macmillan Press Ltd 1983 *Kosta Tispis Directed Energy Weapons,* p26.

27 Dr. Richard M Roberds *"Introducing the Particle beam weapon"* pp3-5 < *www.airpower.maxwell.af.mi./airchroniocles/aureview/1984 jul-aug/roberds.htm*>

28 Ibid pp 6-9.

29 Carlo Kopp " The Electromagnetic Bomb – A Weapon of Electrical Mass Destruction"<*http://www.cs.monash.edu.au/carlo/* >

30 Marie Hewish "Beam Weapon's Revolution" *Jane's International Defense Review,* August 2000. p34.

31 Ibid

32 Ibid p35-36.

33 Ibid pp36-37.

34 Ibid. p37 " the ABL has passed its critical design review in April. The PDRR phase includes the installation of a half power (14 module) laser in a 747-400 test aircraft, which began an 18 month modification program in January, 2002. Installation of ABL system components is due to start at Edward Air Force Base, California, in Jul 2001."

[35] Ibid.

[36] Ibid.

[37] Ibid p38-40 "... High Energy Laser System Test Facility (HELSTF) at White Sands Missile Range in New Mexico integrate the SSL at HELSTF for testing and evaluation and demonstrated the effectiveness of lasers against short-range rockets."

[38] Ibid p39 "... pulsed power includes the design and construction of compact Marx Banks, together with explosive magnetohydrodynamic and maganetocumulative generators.......explosive flux compression generators could typically provide up to 100 MJ and 100 MA, using up to 450 kg of high explosives per shot".

[39] Kapil Kak, "Revolution in Military Affairs", *Strategic Analysis,* IDSA April 2000.

8

Space Application And Indian Space Programme

Space programme has two distinctive paths—the civilian and the military programme. The civilian programme is driven by the national economic and welfare requirements but the military programme is characterised by the technological advances which could enhance military capability and related national security. It is now very clear that space is used for surveillance, reconnaissance, communication, navigation, meteorology and geodesy. Communication satellites have revolutionised both civilian and military sectors by creating facilities to establish contact at the remotest corner of the earth at a very short notice. These satellites perform their missions in many types of orbits, from low earth orbit (LEO) to the highly inclined, eccentric Molniya orbits used by the Russian Federation. One of the important classes of these satellites is the geosynchronous orbit.

The four major orbital paths[1] of the military satellites are shown in the following figures:

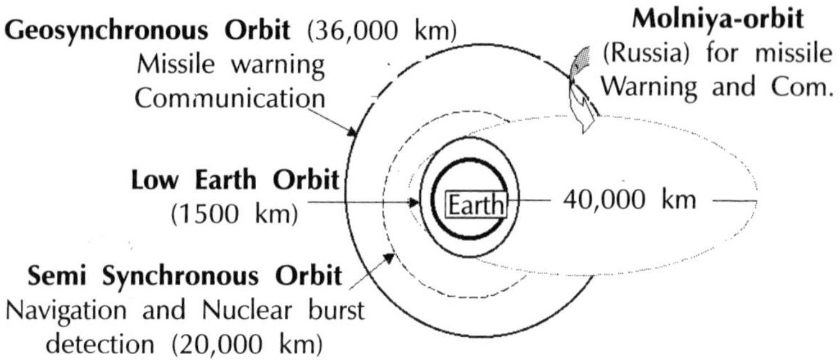

Fig 1. **Orbit Patterns of the Military Satellites**

Military uses of space today include a substantial fraction of benign missions but in the future it may not be the same. Many of the future uses of space will be threatening. And no doubt such development will evoke strong responses from the other nations.

The satellite based High Energy Laser (HEL) weapons system could be a strong probability as reported in the Jane's International Defence Review August 2000 issue[2]. This is a clear indicator of militarisation of the space apart from surveillance reconnaissance or navigation.

In view of such developments it would be in tune to read and get familiarised with few space related common terminologies[3].

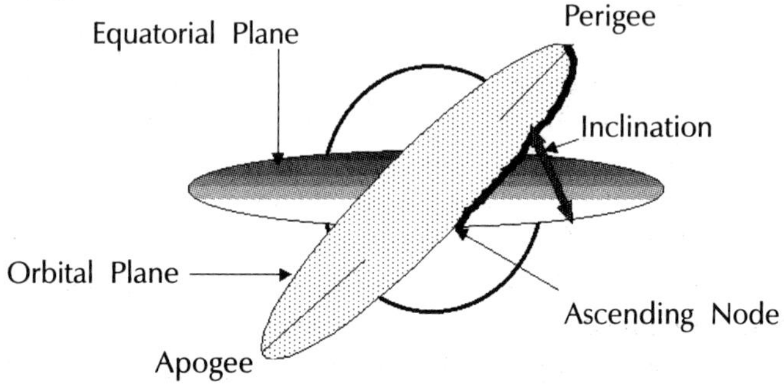

Fig 2. **Orbital Elements**

Perigee: It is the point in an orbit closest to earth, the point where the satellite is moving at its fastest rate.

Apogee: The point in orbit farthest from earth. Also the point in the orbit where the satellite is moving at its slowest rate.

Period: The amount of time it takes for a satellite to go from a particular point on its orbit back to the same point on the next revolution.

Prograde Orbit: An orbit in which a component of the satellite's motion is in a Easterly direction (with the motion of the earth).

Retrograde Orbit: An orbit in which a component of the satellite's motion is in a Westerly direction.

Ascending Node: The point in space where the satellite's orbital plane intersects the equatorial plane on a South to North pass of the satellite. The descending node represents the point of intersection on a North to South pass of the satellite.

Semi-Major Axis: This is half the distance from perigee and apogee. To determine semi-major axis correctly, draw a line inside an ellipse so that the line bisects it. The semi-major axis is equal to one half of the line. This defines the size of the ellipse.

Eccentricity: The amount of orbit deviates from circular . It describes the shape of orbit.

Right Ascension of the Ascending Node: The angular measurement made from the first point of Aries (vernal equinox) measured eastward along the orbital plane to the ascending node. This orients the orbital plane with the rest of the universe.

Argument of Perigee: An angular measurement made from the ascending node, in the direction of the satellite's motion, to the point of perigee on the orbital plane to the ascending node. This orients the orbit within the orbital plane.

Perigee Precision: The location of the perigee shifts along the plane of the orbit due to the earth's irregular shape. This is indicative that a polar orbit's perigee will move from above the North pole southward to over the South pole and back again the rate of perigee precision can be calculated.

True Anomaly: An angular measurement from the point of perigee to the position of the satellite in the direction of the satellite's motion.

Inclination: An angular measurement from the equatorial plane to the orbital plane (measured in a counter clockwise direction around the ascending node).

Epoch Time: The reference time for the parameters of an element set.

Drag: With increasing altitude the atmosphere changes its composition from nitrogen and oxygen mix into a plasmasphere consisting primarily of ionized hydrogen atoms. The atmosphere is very thin and rarefied but the constant collision of a satellite with what few air molecules there are serves to gradually slow down the satellite and thus perturb into a different orbit.

Neutral Drag[4]: In low earth orbits around 1000 km above the earth the atmosphere primarily consists of hydrogen, helium and oxygen atoms. And these are non ionized and therefore do not carry any electrical charge. The drag experienced by the satellite is only due to kinetic collisions. As there are no electrical charges present this type of drag is known as neutral drag.

Charged Drag and Thermal Drag[5]: At a very large height above the surface of the earth the atmosphere becomes the plasmasphere. And it has been noticed that the metal surface of the satellite gets electrically charged and draws charged molecules to it. This electrical interaction causes charged drag. Whereas, thermal drag occurs when the sunlit side of the satellite emits infrared radiation while the shadowed side of the satellite being cooler emits far less radiation, while the other half is not. That creates net thrust way from the sun, which is called thermal drag.

Rotational Speed of Earth[6]: The speed is 0 mile per hour at poles and increases to 1037 miles per hour at the equator. The earth rotates from west to east. As such any launch in a prograde (Easterly) direction receives additional advantages of the earth's rotational speed. Any retrograde launch must overcome the earth's eastward velocity to achieve orbital path. Therefore the greatest advantage can be taken by launching from the equator in prograde direction. To set out the mission plan, careful planning is required to select launch site and azimuth.

Orbital Mechanics: The determination of orbits and trajectories of man made objects in space.

Satellite Ground Traces

Satellite ground traces[7] are required to be understood because it would be an important factor to note what is the path being followed by the satellite in relation to the surface of the earth. This satellite ground trace is formed by the intersection of the surface of the earth and the plane of the satellite. But it must be understood that satellite orbit in space is fixed and it is the earth which rotates under its orbit. As such each orbit it traces out a path on the surface of the earth which is known as ground trace. These ground traces will give out the exact time when the satellite will be visible on a particular part of the earth, and when will it pass over a given area of interest. Also if a ground trace of a foreign satellite is plotted, the purpose of its flight over a particular territory can be determined. Also limitations of a particular mission orbit could be understood by these ground traces.

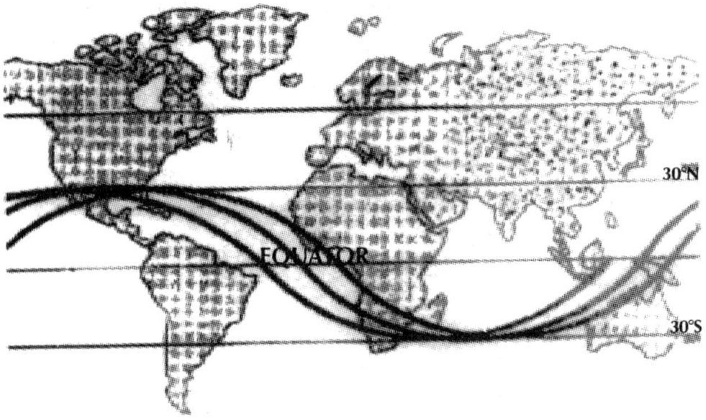

Fig 3. **This is an example of three passes of one satellite on the earth's surface.**

Earth's rotation causes each successive trace of a satellite in a low earth orbit to cross the equator at a point west of the previous path. As a result the ground trace shifts westward after each pass by certain amount of degrees. Therefore, a satellite covers a band around the earth between Southern and Northern hemisphere depending upon its inclination in respect to the equator.

Inclination and its Effects on Ground Trace: Inclination has its own peculiarities. Take a case of three satellite whose inclinations are at 30, 45, and 60 degree respectively. While drawing the ground trace of all of them it would be very clear that they would trace different path on the face of the earth as shown in the following diagram. Path C is 30 deg path B is 45 deg and path A is 60 deg inclination. If the inclination is increased to 90 deg or thereabout the orbit will be called as polar. In polar orbit, a satellite in low earth orbit covers the earth's surface in a single day. Polar orbit satellites are used for surveillance and weather mapping. It won't be out of place to mention here that if a country can place a number of satellites in LEO then it could have uninterrupted surveillance of an enemy target throughout.

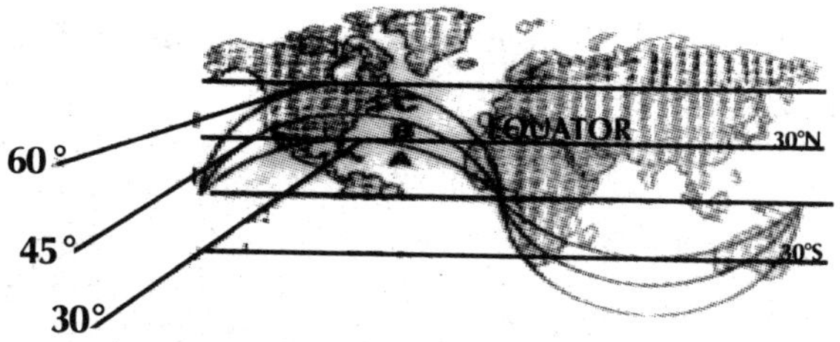

Fig 4. **Satellites with Different Degrees of Inclination**

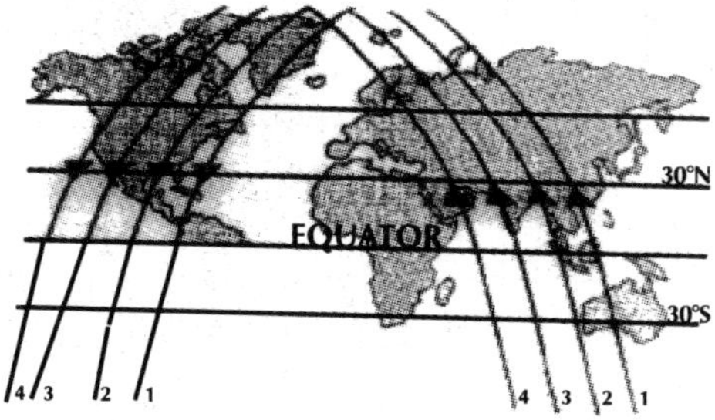

Fig 5. **Polar Orbit of a Satellite**

Geosynchronous Orbit: For Geostationary Orbit[8], synchronous revolution of the satellite with the rotation of the earth is the criteria. Any satellite in that orbit will appear to hover stationery over a point on the earth's surface. Satellites are placed at an approximate altitude of 19,300nm above the surface of the earth. The footprint of the satellite remains constant. The orbit is circular and equatorial with inclination 0 degree and the satellite will have its footprint over an arc of about 42 percent of the earth's surface. It should be understood that it is not possible to orbit a satellite which is stationary over a point and not on the equator. As the satellite remains more or less stationary it is ideal for communication purposes. Limitations of the geostationary families are being the space for the satellites along the belt so limited that frequencies allocated to this purpose do not result in interference between satellites on uplink or downlink. Since the geostationary orbital plane is not coincident with the plane of the earth's orbit or that of the moon's orbit, the gravitational attraction of the sun and the moon act to pull the satellite out of its equatorial orbit, gradually increasing each satellite's orbital inclination. In addition, non circular shape of the equator causes these satellites to be slowly drawn to one of two stable equilibrium points gradually increasing each, satellite's orbital inclination and goes out of their stationary position. To correct these perturbations sufficient fuel is loaded into all geostationary satellites to periodically correct any changes that take place over the entire lifetime of the satellite. These periodic corrections are known as station keeping. Once the satellite has exhausted its fuel, its inclinations will begin to grow and it will begin to drift in longitude and may present a threat to other geostationary satellites. At times, these satellites are boosted to slightly higher orbit at the end of their planned lifetime to prevent them becoming a menace to other satellites at that altitude.

Launch Window [9]: The term launch window is associated with the period of time for specific launch site and desired orbit. Most fuel-efficient launch can be done during the period. The period of the window is determined by the launch vehicle's performance. If the latitude of the launch site is lower than the inclination then there will be two launch windows a day. But if the latitude of the launch site is higher than the inclination then the launch window will be only one a day. It is rather difficult to directly launch a satellite into an orbit whose inclination is less than the latitudes of the launch site.

Maneuvers in Orbit

The purpose of an orbit maneuver is to change one or more parameters of the orbital element set. Basically an orbit of satellite is stable. This stability is an advantage yet at times it becomes necessary to change various parameters to suite the operational requirements. Space supply, rendezvous, interception needs may call for the change of the orbital elements at crucial moments.

There are two types of maneuvering, in plane and out of plane. In plane maneuvering means the changing the size and shape of an orbit by a process known as Hohmann transfer[10]. This method requires minimum energy to perform the desired changes in the plane. But the disadvantage of this maneuvering is that it takes maximum time between the two orbits and may at times take couple of hours. However, if operational requirement arises, the transfer can be carried out in much faster manner, which is known as high energy transfer[11]. Out of plane maneuver is needed to change the orbital plane. It is important to know that the plane change is carried out at a point common to both orbits and must be completed at the ascending or descending node.

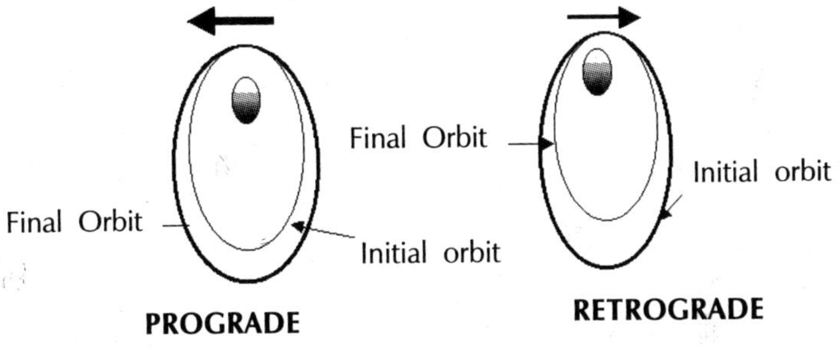

Fig 6. **In plane maneuver of Prograde and Retrograde Satellites**

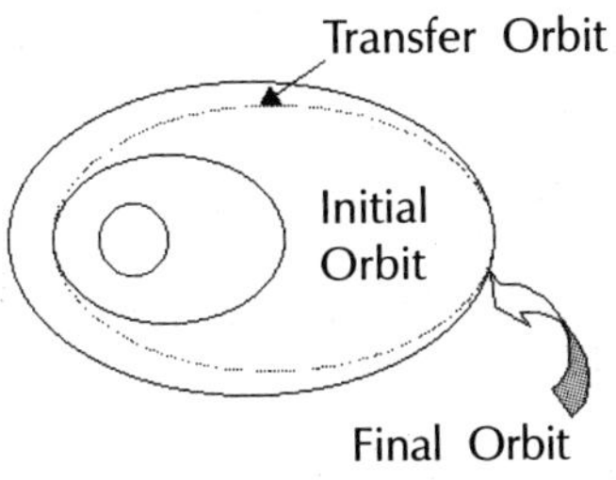

Fig 7. **Hohmann transfer. This method requires minimum energy.**

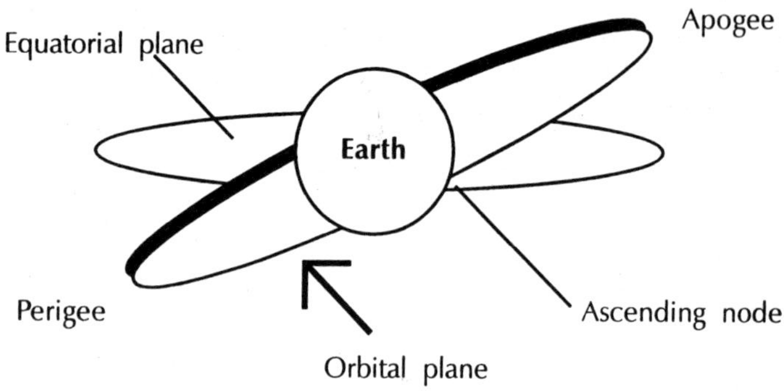

Fig 8. **Change of plane. Generally all changes are made at the nodes.**

An Outside Force known as Peturbations[12]: Satellites in eccentric orbits with very high perigees are particularly susceptible to being perturbed by the gravity of the moon, sun or both. Perturbations are deviations from the normal or expected orbit created by various outside forces. These perturbations can gradually increase eccentricity, which causes perigee height to lower and experience increased drag. Normally to predict a satellite's future position is not difficult because of its more or less rigid path being set by orbital element set. Perturbations result from atmospheric drag, the shape of the earth, solar activity, the magnetic field of the earth, etc. These forces have direct impact on satellite operations and if remain unchecked the satellite could drift to a different orbit than desired.

Global Positioning System (GPS)

GPS[13] may be the only military system used both by military and the civilian sector extensively but exclusively manned by the military. NAVSTAR represents a navigation system heretofore unavailable anywhere. The use of the system has become so overwhelmingly common that from a fighter pilot to an infantry man is carrying the GPS receiver for a ground fix. Lets us understand the system and its impact. GPS provides three dimensional positioning anywhere in the world whether in air or on ground. In terms of accuracy the system provides Spherical Error Probables (SEP) of 16 mtrs. To satisfy military navigation requirements and yet provide other users with positioning capability, GPS provide two levels of service. Precise Positioning Service (PPS) provides the most accurate positioning information (16 mtr SEP) but users have to be authorized by US DoD. PPS capable GPS receiver sets have the ability to receive P (Precise) code signals transmitted by the GPS satellites, and remove the effects of Selective Availability (SA). However, all GPS users have the access to the Standard Positioning Service (SPS). The positioning accuracy provided by SPS is 100 mtrs a circular error probable (CEP), which means this service will place users within 50 mtrs from their true position. This reduced accuracy compared to PPS is because SPS users utilise the C/A (Course/Acquisition) code transmitted by the satellites. This code provides less accuracy but is easier to decode than the P-Code.

Data from at least four GPS satellites must be received for a user set to calculate its three dimensional position. The receiver set calculates the distance to each of the four satellites using the P or C/A code, solves for the user's position via triangulation. The unique combination of delay times will be true for only one location on the earth.

In addition GPS also provides the user with velocity and time data. Velocity is accurate to 0.1 mtr/sec and time to 0.1 microseconds. Velocity data is provided by monitoring the shift in frequency of the navigation signals caused by the velocity of the user. This process works on the principle of the Doppler radar. On the other hand the time information is sometimes considered a mission of its own and is termed as time transfer. Time transfer allows the user to obtain the International Standard of Time to within 100 billionth of a second. The international standard is called Universal Time Coordinated or UTC.

Twenty-four GPS satellites are used to provide nearly continuous three-dimensional coverage worldwide. The satellites are in circular orbit just under 11.000nm above the earth's surface. There are three or four satellites in each of six orbital planes, each plane inclined 55 degrees to the equatorial plane and separated by 60 degrees in right ascension.

Analysis: It may be interesting to note that for the first three decades of the space age, the superpowers found it technically and economically sound to use space for reconnaissance and surveillance, communications, navigation, meteorology and geodesy. But militarisation of the space was also going on at a fairly decent pace, till such time the bubble of cold war burst and the Soviet Union gave away the sole proprietorship of the space to the USA. The USA is making the full use of the situation. GPS has become a daily necessity in both civilian sector and militaries of the world. The development in relation to GPS may not augur well for the countries, mainly the fighting forces. The capabilities of the third generation GPS satellites will have many more capabilities that will be undesirable in military context for other nations. Embracing GPS without reservations would be disastrous for any military. The moderated GPS signal will dictate disaster for any ground or air military operations. Despite protests from all quarters, it is just a matter of time the NMD will be stationed on space-based satellites. Like many others, Indian armed forces have also contributed to this thought of GPS and armed their infantry men and the fighter pilots with the compatible equipment for them to fight efficiently little realising that once the third generation satellites are positioned at the Semi Synchronous orbit, this potent force will be fighting a battle dictated by some terrestrial master.

India in Space

Despite limited resources, India has and is continuing to develop a broad based space programme with indigenous launch vehicles, satellites, control facilities and data processing. Since its first satellite was orbited by the USSR in 1975 and its first domestic space launch was conducted in 1980, India has become a true space fairing country and could take definite lead if compared to many developed nations of the European continent. Present generation Indian remote sensing, communication and meteorological satellites are comparable to many space systems produced by the USA and Russia or France. And as a matter of pride, in a very short period from now India would be among the select countries with capabilities to launch Geostationary launch capability.

The Indian Space Research Organisation (ISRO) was founded in 1969 and is currently under the Department of Space, with its headquarters at Bangalore with a work force of about 17,000. The corporate headquaters are in Bangalore, but activities related to satellites, launch vehicles and applications are carried out at numerous centers throughout the country. The development of the sensors and payloads is carried out at ISRO Satellite Application Center (SAC) in Ahmedabad. ISRO Satellite Centre (ISAC) in Bangalore is responsible for the design, development, and assembly and testing of satellites. At Thiruvananthpuram, Vikram Sarabhai Space Centre (VSSC) looks after the launch vehicle. The Liquid propulsion modules, including cryogenic engines, are developed at the Liquid Propulsion Centre near VSSC. Launching of the satellites is carried from Sriharikota, north of Madras. Master Control facilities for satellite station keeping are located at Hassan near Bangalore. The reception and processing facilities for remote sensing data are available at National Remote Sensing Agency (NRSA) in Hyderabad.

History of Indian Space Programme[14]

Satellite Name	Launcher	Name of the Country	About the Satellite
Apple	Ariane	France	Apple (Arien Passenger Payload Experiment) was launched on 19 June, 1981, from Kourou. This was India's first indigenous GEO test communication satellite. Indian TV programmes and educational teleconferences were relayed, even though a solar sail failed to deploy. The satellite was abandoned or deactivated in Sep 1983.
Aryabhata	SL 8	Russia	Aryabhata was launched on 19 April, 1975, from Kapustin Yar, with Perigee/ Apogee at 563 x 619 km. Aryabhata was the famous Indian mathematician. This was to carry out X-ray astronomy. Indian scientists could not take over the control as the Indian-designed transformer failed and only days of useful life was obtained.
Bhaskara 1	SL 8	Russia	Launched from Kapustin Yar on 7 June, 1979. Bhaskara was a 7th century astronomer. It was loaded with two TV cameras and microwave radiometers, Useful ocean and land surface data were received, but the cameras had malfunctioned. The satellite reentered orbit on 17 February, 1989

Bhaskara 2	SL 8	Russia	Again launched in 1981 from Kapustin Yar. First achievement was that 300 television images of Indian subcontinent were received and the satellite declared operational. The telemetry was still being received until 1991. The satellite re-entered orbit on 30 November 1991.
Rohini-1B	SLV 3	India	From Sriharikota launched on 18 July, 1980. It was India's first successful launch. The Rohini-1B followed the failure of the Rohini 1A, on 10 August, 1979.
Rohini-2	SLV 3	India	Rohini 2 was launched on 31 May, 1981, by SLV-3 from Sriharikota. For the first time it was given the imaging capability and carried an imaging system, but the orbit failed to achieve its planned orbital path of 296 x 834 km and re-entered the earth orbit only after eight days.
Rohini-3	SLV	India	Launched on 17 April, 1983, from Sriharikota. Weight 41.5kg. Orbit - 371 x 861 km, 46.6°. It carried 2 cameras and L-band beacon. It returned 5000 earth images and was activated on 24 September, 1984. On 19 April, 1990, re-entered earth orbit.

SROSS 1 **SROSS 2** **SROSS C2** **SROSS C3**	ASLV ASLV PSLV D2 ASLV D3		Stretched Rohini I was launched on 24 March, 1987. But was lost in vehicle failure. SROSS 2 was launched on 13 July, 1988, but again it was lost in the failure of second ASLV. SROSS -C 2 launched on 4 May, 1994 SROSS C3 launched on 20 May, 1992. The satellite carried a National Physics Lab's Retarding Potential Analyzer for measuring low altitude ionosphere and thermosphere and 20-3000 KeV Gamma ray Burst experiment.

Remote Sensing Satellites of India[15]

IRS-1A	SL-3 Vostok	Russia	India really joined the space club once the IRS-1A, the first remote sensing satellite, became operational riding on the Russian Vostok rocket. This was launched on 17 March, 1988, from Tyuratum. A four band imaging system with CCD cameras returned overlapping 73 meters and 36.5 meters resolution images from sun synchronous orbit..

IRS -1C	Molniya - M	Russia	From Baikanour this was launched on 28 December, 1995, from Baikanur into a retrograde (99° inclination) orbit, 816 / 818km (sun - synchronous)

IRS -1D	PSLV-C1	India	IRS-1D was launched from Sriharikota in September 1997, using PSLV. IRS-1D was placed in an elliptical orbit, The satellite imagery will be used to map and also monitor calamities.

IRS-1E	PSLV-D1	India	Another remote sensing satellite was launched on 20 September, 1993, from Sriharikota and it was this satellite IRS-1E that carried the Indian LISS-1 camera,the German MEOSS stereo scanner and an earth radiance monitor in the CO2 band. The PSLV control software prevented attaining orbit.
IRS-P2	PSLV-D2	India	The IRS-P2 satellite has completed more than six years of successful operation in orbit, launched on 15 October, 1994, from Sriharikota.
IRS-P3	PSLV-D3	India	This was the third test launch of the Indian PSLV on 21 March, 1996, from Sriharikota. Perigee/ Apogee 802/848 km, inclination 98.8°
IRS-P4	PSLV-C2	India	The IRS-P4, launched on 26 May, 1999, also known as the 'Oceansat-1', has a multi-frequency radiometer and will be used for monitoring the ocean surface.
IRS-P5	PSLV	India	Defence forces must be waiting for this satellite to be launched shortly because this has the promise to have a pan camera with a resolution of 1-2.5 metres. It would be mainly useful for map making and terrain modeling. And that is why it is known as Cartosat. Date of launch is not yet known.
IRS-P6	PSLV	India	Another advanced version of the Cartosat and it is expected that IRS-P6, also known as the Resourcesat', will have even had a higher resolution multi-spectral camera. and will provide better facilities than its predecessors. Should be launched in 2002.

Indian Communication Satellites[16]

INSAT-1A	Delta 3910 Pam	India	Launched on 10 April, 1982, and abandoned on 4 September, 1984. The launch was from Cape Canaveral. With a weight of 1152 kg. A jammed C-band antenna was released 12 days after the launch I.e. on 22 April, 1982. In addition to providing nationwide TV to rural areas, it returned 30 minute meteorological images. Abandoned in September 1982 short of its 7-year life. The reported insurance payout was about $70 million.
INSAT-1B	STS-8	USA	One of the most successful satellites in the field of communication launched on 30 August, 1983, from Cape Canaveral. Weight was 1152 kg at launch. It almost suffered the same fate as its predecessor. Video recordings have suggested, that it might have been struck by orbiter debris during release, though this could not be confirmed. Full operation was achieved in October 1983. It continued operating satisfactory into 1990 with all its 4375 two-way voice or equivalent circuits in use; it was only when the INSAT 1D was launched that it was relegated to spare status. Around 36,000 earth images were returned. Eleven of the 12 C-band and two S-band transponders provided communications and direct nationwide TV services to thousands of remote villages, plus a comprehensive weather and disaster-warning service. Total cost of the INSAT-1B and the backup INSAT-1C, including the PAM-D launch was estimated at $ 140 miliion.

INSAT-1C	Ariane 3	France	Launched on 21 July, 1988, from Kourou (launch 1190, GEO 700 kg) for location at 93.5°E to bring INSAT system up to full capacity. Earth lock was lost on 22 November, 1989, and it was abandoned. Half of the 12 C-band & two S-band transponders were lost when a power system failure knocked out one of the two fuses, but the meteorological Earth images and data collection system were both fully operational.
INSAT-1D	Delta 6925	USA	Launched on 12 June, 1990, from Cape Canaveral to conclude the first generation INSAT series. Launch was planned for 29 June, 1989, but 10 days before, it was seriously damaged following installation on the Delta rocket when a 34 kg crane hook fell directly on to it. Ford Aerospace repaired the fully insured satellite at a reported cost of $10 million. It also suffered $150,000 of damage during the 1989 California earthquake. The INSAT-1D replaced INSAT-1B on 17 July, 1990.
INSAT-2A	Ariane 4	France	Launched on 09 July, 1992, from Kourou for location at 74°E. Commissioned on 06 Aug Earth lock was lost on 22 November, 1989, and it was abandoned in 1992. Carried 18 C-band and two S-band transponders, plus a 2 km resolution IR, meteorological imaging system.
INSAT-2B	Ariane 4	France	Launched on 22 July, 1993, from Kourou and was declared operational at 93.5° East on 10 August, 1993. In April 1999, one on board transponder failed. And in July 2000, the satellite completed its seven-year life span By Nov 2000 it was declared unserviceable
INSAT-2C	Ariane 4	France	The satellite was launched on 06 December, 1995, from Kourou. The INSAT-2C replaced imaging payload with improved telecom services.

INSAT-2D	Ariane 4	France	INSAT-2D was a story of similar failure launched on 4 June, 1997, from Kourou, with a massive weight of 2019 kg. The satellite was to give a big boost to television and telephone services. But due to power failure problem in October 1997, it was abandoned.
INSAT-2E	Ariane 4	France	It was supposed to be launched in October, 1998, but delays caused the launch on 03 April, 1999. This was a successful launch and the satellite is currently in orbit.
INSAT-3A	Ariane 5	France	New generation of the INSAT-series of satellites. This satellite is a hybrid, multi-purpose satellite with a mix of C/Ku band communications and a weather monitoring system.
INSAT-3B	Ariane 5	France	INSAT 3B was successfully launched on 22 March, 2000.
INSAT-3D			Launch date is not known.
INSAT-3E	Ariane 5	France	A dedicated communications satellite with 36 C-band transponders. Launch date is late 2002.

PSLV-C3		India	ISRO's Polar Satellite Launch Vehicle, PSLV-C3, successfully launched three satellites Technology Experiment Satellite (TES) of ISRO, BIRD of Germany and PROBA of Belgium- into their intended orbits on (October 22, 2001.

DBS	GSLV-D1[17]	India	ISRO successfully launched the GSLV-D1 with GSAT on 18 April, 2001. The 401 tonne, 49.1 mtr tall spacecraft carrying 1540 kg experimental satellite lifted off from launch pad at 15:45 pm. The mission heralds a new confidence in ISRO towards the establishment of indigenous capability for launching INSAT class satellites.
2002	**INSAT-3C**	India	Successful launch of INSAT-3C by Ariane from Kourou French Guyana, (January 24, 2002).

Indian Remote Sensing Satellites

It may be found from the above that following the successful flights of Bhaskar 1 and Bhaskar 2, launched in 1979 and 1981respectively, India began developing an indigenous IRS (Indian Remote Sensing Satellite) programme to support the national economy in the areas of agriculture, water resources, forestry, ecology, geology, watersheds, marine fisheries and coastal management. The India Remote Sensing Satellites are the mainstay of National Natural Resources Management System (NNRMS) for which Department of Space is the nodal agency providing operational remote sensing data services. Data from IRS satellites is received and disseminated by several countries all over the world. With the introduction of high-resolution camera new application in the areas of urban sprawl, infrastructural planning and other large-scale applications for mapping have been initiated. NNRMS now cover diverse fields such as crop acreage and yield estimation, drought warning and assessment, flood warning and control, agro-climatic planning, wasteland management, water resources management, underground water exploration, prediction of snow melt run-off, fisheries development, mineral prospecting, forest resources survey. Under this application, an Integrated Mission for Sustainable Development (IMSD) was initiated in 1992. IMSD under which 174 districts have been identified aims at generating locale-specific action plans for sustainable development.

As of late 2000, five IRS satellites were in the orbit and more are scheduled to join this family. The Indian Space Research Organisation (ISRO) and its commercial arm Antrix Corp Ltd successfully launched the IRS-1D the earth Imaging Satellite on 29 September, 1997, from Sriharikota. This is similar to IRS -IC. The dual use of these satellites provides 5.8 mtr resolution images to customers twice as often as was possible with just the IRS-IC. This has serious military implications. The resolution of both IRS-1C and 1D panchromatic (0.50.75 um-black and white) imagery, which resembled five-meter pixel detail. This resolution, which as of early 1998, was the best of any civilian remote sensing satellites in the world, is superior to the 8 mtr resolution initially reported for the panchromatic imager. These satellities are also equipped with two wide Field Sensors (WiFS) that cover a 774 square kilometer (481 square mile) area in a single image, as well as LISS-3 4 band (0.52-0.59, 0.62-0.68, 0.77-0.86 and 1.55-1.70 um) multispectral sensors that provide 23.5-meter resolution. However it can be pointed off the orbit, which allows two to four days revisits to specific sites. It is expected that the new IRS 2A and IRS 2B will be with new LISS-4 sensor suite.

In a press release in August 2001, ISRO announced that IRS-IB, which was launced on board the Russian Vostok launcher on August 29, 1991, has successfully completed 10 years of operation. The 990 kg satellite carries two Linear Imaging Self Scanner (LISS-1) with a spatial resolution of 72.5 m and ground swath of 148 km and LISS-IIA and LISS-IIB. The satellite is placed in a polar sun synchronous orbit at a height of 904 km with an orbital period of 103 minutes. The satellite can image the entire earth once in 22 days

IRS-P4 (OCEANSAT-1): It has payload of Ocean Colour Monitor (OCM) which is a solid state camera operating in push broom scanning mode, using linear array Charged Coupled

Devices as detectors. This camera has 8 narrow spectral bands operating in visible and near infrared bands with a spatial resolution of 360 mtrs, 12 bit radiometric quantisation and swath at 1420 km. OCM has separate wide angle, telecentric optics and CCD detectors for each spectral band, which helps to optimize the performance of individual channels. As ISRO explains, that since OCM is designed to provide the highest spatial resolution among the contemporary ocean colour sensors with unique regional and global coverage, it is expected to provide ample opportunities/ application potentials, for both scientific and commercial and military uses. The OCM data requires the ground station to be equipped with X-Band. Currently, the Indian ground station at NRSA Hyderabad is receiving data in real time. India is trying to build up a global network of ground stations covering other regions of the world. It is expected that these international ground stations will provide the needed geometrically and radiometrically corrected data to the selected places under some special arrangements being worked out with the ground stations.

OCEANSAT-1 also carries a Multi-frequency Scanning Microwave Radiometer (MSMR). MSMR with its all weather capability provides measurement in four frequencies with dual polarisations. A conical scanning system of an offset parabolic reflector provides a constant look angle of 43.13° at each scan position. MSMR data is useful for deriving both physical oceanographic and atmospheric parameters. The specific advantage of MSMR is that it has the frequency channel at 6.6 Ghz (V & H) which is not available in any other contemporary satellites. MSMR data for the whole globe is covered in 48 hours. Global data is recorded on an onboard solid state recorder continuously and downloaded only at National Remote Sensing Agency (NRSA) ground station at Hyderbad.

Indian Launch Vehicles[18]

Indian Launch Vehicles

Vehicle	Launch Dates	Result
1. **SLV-3 E1**	August 10, 1979	A jammed valve in the second stage control system resulted in the leak of oxidiser.
2. **SLV-3E2**	July 18, 1980	Successful
3. **SLV-3D1**	May 31, 1981	Successful
4. **SLV-3D2**	April 17, 1983	Successful
5. **ASLV-D1**	March 24, 1987	Unsuccessful due to non -ignition of first stage.
6. **ASLV-D2**	July 13, 1988	Unsuccessful. The flight was normal only upto 46 seconds after lift off.
7. **ASLV-D3**	May 20, 1992	Successful
8. **ASLV-D4**	May 4, 1994	Successful
9. **PSLV-D1**	September 20, 1993	Unsuccessful due to software error in onboard guidance and control processor.
10. **PSLV-D2**	October 15, 1994	Successful
11. **PSLV-D3**	March 21, 1996	Successful
12. **PSLV-C1**	September 29, 1997	Successful
13. **PSLV-C2**	May 26, 1999	Successful
14. **GSLV-D1**	April 18, 2001	Successful

Polar Satellite Launch Vehicle (PSLV)[19]: PSLV was developed by ISRO to place 1000 kg class Indian remote sensing satellite into Polar Sun Synchronous Orbit (SSO). First successful flight was conducted in October 1994. Salient features of PSLV- 3 are given as follows:

- Overall length : 44.4 m

  Lift-off weight : 294 t

  No. of stages : 4

  Payload : TES, BIRD, PROBA

  Orbit (SSO) : 570 km

The first successful flight was conducted in October 1994, and the capability of PSLV had been enhanced from 805 kg to 1200 kg into 820 km SSO. PSLV also has the capability to launch 3,500-kg satellites into 400 km low earth orbit and 1000 kg satellite into Geo-synchronous Transfer Orbit.

In its present configuration, PSLV has four stages using solid and liquid propulsion system. The **first stage** is one of the largest solid propellant boosters in the world and carries 138 tonne of Hydroxyl Terminated Poly Butadience (HTPB) propellant. It has a diameter of 2.8 m. The motor case is made of maraging steel. The booster develops a maximum thrust of about 4, 430 kN. Six strap-on motors, four of which are ignited on the ground, augment the first stage thrust. Each of these solid propellant strap-on motors, carries nine tonne of HTPB propellant and produces 667 kN thrust.

The **second stage** employs indigenously manufactured Vikas engine and carries 40 tonne of liquid propellant- Unsymmmetrical Di-Methyl Hydrazine (UDMH) as fuel and Nitrogen tetroxide (N204) as oxidiser. It generates a maximum thrust of about 724 kN.

The **third stage** uses 7 tonne of HTPB-based solid propellant and produces a maximum thrust of 324 kN. Its motor case is made of polyaramide (Kevlar) fibre.

The **fourth** and the **terminal stage** of PSLV has a twin engine configuration using liquid propellant. With a propellant loading of 2 tonne (Mono-methyl hydrazine and Mixed Oxides of Nitrogen), each of these engines generates a maximum thrust of 7.4 kN.

The 3.2 m diameter metallic bulbous heat-shield of PSLV, is of isogrid construction and protects the spacecraft during the atmospheric regime of the flight. PSLV control system includes:

- First stage : Secondary Injection Thrust Vector Control (SITVC) for pitch and yaw, reaction control thrusters for roll and SITVC in two strap-on motors for roll control augmentation.
- Second stage : Engine gimbal for pitch and yaw and, hot gas reaction control for roll.
- Third stage : Flex nozzle for pitch and yaw and PS-4 RCS for roll.
- Fourth stage : Engine gimbal for pitch, yaw and roll and, on-off RCS for control during the coast phase.
- The inertial navigation system in the equipment bay, which is located on top of the fourth stage, guides the vehicle from the vehicle from lift-off to spacecraft injection into orbit. The vehicle is provided with instrumentation to monitor the vehicle performance during the flight. S-band PCM telemetry and C-band transponders cater to this requirement. The tracking system provides real-time information for flight safety and for preliminary orbit determination once the satellite is injected into orbit.

PSLV employs a large number of stage auxillary systems for stage separation, heat-shield separation and jettisoning, etc.

PSLV-C3 Flight Profile: The flight profile for PSLV-C3 mission had been modified to place the three satellites in their specified orbits—the Indian satellite (Technology Experiment Satellite) and the German satellit Bispectral and Infrared Detection (BIRD) in a 568 km sun-synchronous orbit and the third satellite, the Belgian Project for On Board Autonomy (PROBA) in a 568 x 638 km elliptical orbit - which require firing the reaction control thrusters of PSLV fourth stage for about 500 seconds after the separation of first two satellites. A Data Storage Unit (DSU), which stores the telemetry parameters and transmits in delayed mode had been introduced to take care of any break in the radio visibility between the vehicle and the ground stations located at Thiruvananthapuram and Mauritius as well as recording of the PROBA separation event. The PROBA separation related parameters were transmitted when the stage passed over the Lucknow ground station of ISRO.

With the successful launch of three satellites simultaneously for the second time, ISRO's PSLV has proved itself as a promising vehicle with prospects for commercial launching of satellites even while serving its primary goal of launching Indian remote sensing satellites. PSLV is also proposed to be used for a geo-synchronous mission for launching the ISRO's METSAT. Thus, PSLV is poised to become a versatile launch vehicle in the coming years and, along with GSLV, make India self-supporting for its launch services in the near future. Brief details of the three satellites launched are given below:

- **Technology Experiment Satellite (TES),** weighing 1108 kg, is an experimental satellite to demonstrate and validate in orbit, technologies that could be used in the future satellites of ISRO. Some of the technologies that are planned to be demonstrated in TES are solid state recorder, X-band phased array antenna, improved satellite positioning system, miniaturised TTC and power system and, two-mirror-on-axis camera optics. TES will also carry a panchromatic camera for remote sensing experiments. Soon after its injection into orbit, the solar panels on board TES were deployed automatically to generate necessary electrical power for the satellite. Further in a series of operations conducted from the ISRO Telemetry, Tracking and Command (ISTRAC) stations at Bears Lake in Russia and Lucknow in India, the satellite was put in three axis stabilished mode using the reaction wheels on board the satellite. The satellite health is being continuously monitored from the Spacecraft Control Centre at Bangalore along with the ISTRAC network of stations at Bangalore, Lucknow, Mauritius, Bears Lake and Hyderabad.

- **PROBA** (Project for On Board Autonomy) is a small satellite of Verhaert, Belgium, weighing 94 kg. The payloads in the satellite include high resolution camera with 115 mm diameter aperture and wide angle camera having aperture of 60 mm.
- **BIRD** (Bispectral and Infrared Remote Detection) is a small satellite of the German Space agency, DLR, weighing 92 kg. It is intended for testing small satellite technologies and a new generation of infrared sensors for the detection of hot spots like forest fires and volcanoes from space

(PSLV Launch Successful, adopted from ISRO press release October 22, 2001)

Cryogenic Engine[20] : Following the US pressure Russia stopped supplying the cryogenic engines required for the development of GSLVs. As a result ISRO took this as a challenge to develop the same indigenously. The cryogenic engine development has been a challenging task with breakthroughs needed with respect to many disciplines including analytical tools, material characterization, fabrication processes, test facility establishment, handling of cryogenic fluids and finalization of test sequence.

The development of Cryogenic Upper Stage for Geo-synchronous Satellite Launch Vehicle (GSLV): On March 30, 2002, with the successful test firing of the indigenous cryogenic engine for a duration of 12 minutes ISRO has crossed another important milestone. The test was carried out at the Liquid Propulsion System Centre (LPSC) Test complex at Mahendragiri in Tamil Nadu. This hot test was preceded, in the last few weeks, by four shorter duration tests of 10, 40, 40 and 200 seconds.

Future Launch Schedule

2002-2003	2003-2004
IRS-P6	P5 Cartosat-2
INSAT 3E INSAT 3A	INSAT 3D
MET SAT	
GSAT 2 GSAT 3	
PSLV-C4 PSLV-C5	C6 C7
GSLV 02 03	C1

Indian Military in Space: India no doubt has come a long way in space application. Successful launch of the GSLV will place India in the select club. However, in the case of space imagery India almost has achieved the best so far. If the number of sun synchronous satellites is increased, the Indian military will be able to reap a rich dividend. Because, our military departments do not have any independent space capability or do not have a defence satellite of their own. For satellite communication, image sensing, battle damage assessment, Indian military will have to depend entirely on ISRO. No doubt the defence departments must cater for such dependence in their war making capabilities. The visit time of each image sensing satellite over a particular area of interest will be the major guiding factor to understand the course of the air land battle and to carry out effective battle damage assessment. To this effect Defence Image Processing and Assessment Centre (DIPAC) has been established as a tri-service institution. DIPAC is liaising closely with ISRO for satellite availability in defence application. Each service HQs has its own space cell, which in turn gets its space imagery and other associated information from DIPAC. Effectiveness of DIPAC has not been tested in any war scenario in the past. As such the services headquaters will have to work out a plan to get information of the battlefield in real time from DIPAC for better effectiveness.

The successful launch of GSLV on 18 April, 2001, has heralded a new chapter of Indian space community. India becomes one of the five countries that now have the capacity to launch geosynchronous satellite. This means that Indian INSAT class satellites will be placed in the correct orbit in future at a much lesser cost. *Times of India* pointed out in their 19 April, 2001, issue that Indian space programme had been totally civil oriented. India now needs an autonomous space launch capability for military reasons for surveillance and communication, but not for offensive purpose which is banned.

But space has become an integral part of aerospace power. Air power cannot be exercised in isolation. Indian achievements have been impressive and the time is ripe for it to harness these achievements for military purposes as well. It may be pointed out that from the very onset of space adventurism, both Russia and the USA understood the importance of this fourth dimension to conduct warfare and began militarisation of space. And it was a declared policy of the USA to project its power through space in any future combats, which has been displayed adequately during the Gulf, Kosovo and recently concluded Afghan war. Even China has taken strong steps to exploit space for military purposes. Unless Indian military understand these developments and make thrust towards militarisation of space and take steps to claim its rightful place, it would have to take a back seat in the future power projection through this fourth dimension.

Notes:

1 H Guyford and Hieinz R Pageis, "The High Technologies and Reducing the Risk of War, The New York Academy of Sciences, 1986 pp5-10.

2 Marie Hewish "Beam Weapon's Revolution" *Jane's International Defence Review*, August 2000. p34.

3 Capt Tery L. Donelson and Capt. Kimberly M. Walker, *Orbital Mechanics*, Peterson AFB, Air Force Space Command, Space Operations Orientation Course Handbook, July 1993, pp23-25.

4 Visual Satellite Observing, pp6-7. <ftp:/ftp.cc.utexus.edu/people/worden/vshop/FAQ/status>

5 Ibid

6 Ibid

7 Capt Tery L.Donelson and Capt.Kimberly M.Walker, *Orbital Mechanics* Peterson AFB, Air Force Space Command, Space Operations Orientation Course Handbook, July 1993, pp26-28.

8 Dr. TS Kelso, Basics of the Geostationary Orbit, pp3-4. <tskelso@celestek.com>

9 Capt. Tery L. Donelson and Capt. Kimberly M.Walker, *Orbital Mechanics* Peterson AFB, Air Force Space Command Space Operations Orientation Course Handbook, July 1993, pp33-34.

10 Ibid pp 37-39.

11 Ibid.

12 Visual Statellite Observing, pp 8-9. <ftp: /ftp.cc.utexus.edu/people/worden/vshop/FAQ/status>

13 Capt. Tery L Donelson and Capt Kimberly M Walker, *Navstar Global Positioning System* Peterson AFB, Air Force Space Command Space Operations Orientation Course Handbook, July 1993, pp 151-158.

14 Annual Report, Government of India Department of Space 1995-1996

15 Annual Report, Government of India Department of Space 1997-1998

16 Annual Report, Government of India Department of Space 1998-1999

17 Annual Report Government of India Department of Space 1999-2000

18 ISRO Press release dated October 21, 2001.

19 Ibid

20 ISRO press release, March 30, 2002.

9

Space Militarisation:

Concerns For India

Right on the heel of President Bush's National Missile Defense (NMD)plan on May 01,2001, came the special Pentagon briefing of the Secretary of Defence Donald Rumsfeld on May 8,2001. He mentioned very categorically[1], "Space management and organizational changes are a part of our initial efforts to transform the US department and establishment and reform DoD structures, process and organisation. They will help the U.S to focus on meeting the critical security space needs of the 21st century." Timing of the statement is very significant. It clearly defines the intensions of the Bush administration will be taking for weaponisation of space. Being a member of the Space Commission Rumsfeld had been in touch with the space related developments and made up his mind for a long time to make such a pronouncement. This explains, such concerted announcements came as a no surprise to many US watchers. Civilian scientific missions are part of the regular activities and so far there have been 4,000 worldwide orbital space launches. But the question arises, can scientific missions in space be separated from military activities? Or are scientific missions in space a way of justifying and improving military activities? Countries like the USA understood this space game plan, way back in July 1955, when it announced its intention to launch a satellite to study upper atmosphere during the International

Geophysical Year (a worldwide scientific event from 15 July, 1957 to 31 December, 1958). But the Soviets surprised the USA when they launched their first Sputnik in 1957[2]. The US desperately trying to catch up with the Soviets received a jolt when their first launch attempt on 6 December, 1957, blew up on the launch pad with the whole world watching. Despite this humiliation, US Explorer 1 lifted off on 31 January, 1958[3], and was designed to discover Earth's radiation belt. And thus began the race for space, scientific exploration versus the military application. Rumsfeld has been very deep into it and stated very categorically, "Satellites are our worldwide eyes and ears[4]." Having left the erstwhile Soviet block well behind in the race for space, the USA will try to have a complete domination of outer space. And no doubt the main component of the NMD will be satellite-based warning and may be later on satellite-based interception of the incoming warheads. Congress will go along willingly as long as Rumsfeld can convince them it is the best way to ensure the US security. Nothing is better than to tell the Congress that if they do not go along with the plan it will jeopardise the national security. The US Government passed the National Aeronautics and Space Act in July 1958, and National Aeronautics and Space Administration (NASA) began to function from 1 October, 1958[5]. Scientific adviser to President Eisenhower James R Killian played a significant role towards the formation of NASA. Space exploration for only civilian application was much against the desires of the Congress at that time. The US security angle focused the Congress attention to support all space programmes and desired a much stronger military representation in every space programme. The pronouncements of Bush and his Defence Secretary only echo such sentiments. These are in tune with the US perspective of global domination through the medium of space.

Nevertheless, the US Congress went ahead and created the Civil-Military Liaison Committee to coordinate NASA and Department of Defence (DoD) activities and followed by the creation of National Aeronautics and Space Council[6] chaired by the President of the United States. It is during this period that the National Space Policy was spelt out clearly.

Militarisation of Space

The race for the space actually began when the US Navy launched the world's first military navigational satellite system, **Transit 1A**[7] in September 1959, barely after one year of first US space launch. This helped the Polaris submarines to reduce their navigational error to about one mile. The US Army was not very far behind. They launched their first military communication satellite Courier 1B on 4 October, 1960. Satellite assisted navigation and communication were the precursor to all military space activities.

The predecessors of Rumsfeld went ahead to test launch first US Antiballistic Missile (ABM) on 16th December, 1959, and deployed the US Discoverer reconnaissance satellites[8]. In reply the Soviet Union developed several anti-satellite (ASAT) system in the 1960s[9]. This prompted the USA to follow suite. Within three years of the first space flight, traffic jam in space was quite evident because no body believed in any law or common heritage of mankind.

Consolidation by the US military of their space application was quick enough. North American Air Defence Command came into being in Colorado and became operational on 12th September, 1957[10]. NORAD became the primary agency of USA to survive and fight a nuclear war. It kept a continuous vigil for any nuclear attack on the American soil. In October 1960 NORAD was given the task of space defence as well with the formation of the space detection and tracking system. Space surveillance is one of the major missions of NORAD at present. It may be mentioned here that an important component of the missile warning system was and still is the Ballistic Missile Early Warning System (BMEWS)[11], which includes powerful radar systems in Alaska, at Thule in Greenland and at Flying Dales in North Yorkshire.

With such rapid militarisation of space, Rumsfeld statement can be examined more critically. He said, "Our dependence on operations in space however, makes us somewhat vulnerable to new challenges. It's only logical to conclude that we must be attentive to these vulnerabilities and pay careful attention to protecting and promoting our interest in space. ... Our first choice is not to prevail in a conflict, but to be arranged in a way that can dissuade others form engaging in acts hostile to the United States national security interests and therefore deterring conflict from occurring[12]." The statement makes sense when the principal adversary USSR has completely fallen off from the race for space and Russia's present economic predicaments cannot challenge the US aspirations.

But things were different in the late sixties and early seventies. The Soviets manned space programme took the USA by surprise. Yuri Gagarin[13] created history by being the first man to orbit the earth in Vostok 1. Only a few days later on 5 May, 1961, US Navy Commander Alan B Shephard[14] became the first American to go into space and precisely twenty days later President Kennedy's announcement came to put a man on the moon.

Military space application for the first time was tested during the Cuban crisis in 1963. The resolutions could not have been as clear as of today but it was clear enough for the US photo interpreters to recognise Soviet nuclear bluff[15]. The USA stopped USSR ships from entering Cuban waters.

Four years later in 1967 UN General Assembly introduced "The Outer Space Treaty"[16] which reserves space for peaceful purposes. The question that was raised earlier can be raised again, can scientific missions in space be separated from military activities? This treaty bans nuclear weapons and weapons of mass destruction from space. It also states that nations shall not contaminate space and "states shall be liable for damage caused by their space objects." And that "the exploration and use of outer space shall be carried on for the benefit and in the interest of all countries and shall be the province of all mankind[17]." Here the biggest question arises, in post cold war scenario who could be the space marshal? What do we understand by space contamination and what could be the liabilities of the states performing such an act like contamination. When nuclear powered satellite fails to maintain its orbit and re-enters the earth who should be made responsible and to what extent? The sole super power at this juncture is washing its hands off. It wants to install NMD for its ultimate safety from misadventures by the rogue states. And two citizens of that country, Richard Garwin and Hans A Bethe, had described earlier how China or the Soviet Union could easily defeat the US missile defence then under development[18]. But at the same time they retorted that the system could provoke Soviet Union and escalate the arms race. To stem the tide of the arms race both the super powers came together to sign the Anti Ballistic Missile (ABM) Treaty[19] in 1972. And to test each other's strength rather than carry out scientific exploration.

Two years later, Secretary of Defense Harold Brown informed the world that Soviet Union had finally perfected an Anti Satellite (ASAT) system[20], which made Carter administration to change its course and go in for the Space Defense Programme. The entire programme devoted towards research into ASAT technology, satellite survivability and improved space surveillance. This was in 1977. Since then the USA has been spending considerable amount of time and money to this effect and twenty-four years later its aspirations to install NMD may be well understood. In 1981 the USA took a giant leap for perfecting the space system by launching the first space shuttle on 12 April, 1981[21]. Astronauts John Young and Robert Crippen landed Columbia successfully only two days later and became the heroes of space. This unique achievement ushered in certain other factors not known before. President Reagan came out openly to support the Shuttle programme. The space shuttle henceforth became the primary launch system of the US national security space programme. Reagan was the one who told the US space community to take necessary steps for *"assured access to space"*. Followed by it was the DoD directive, which made abundantly clear that the ASAT programme was nothing but to deny the enemy, the use of space and space assets in time of crisis. It also extended the principle of sovereign rights over nation's space assets to include the right to defend those assets in space. Then came the famous Reagan Star War speech on 23 March, 1983, in which he announced the Strategic Defense Initiative (SDI) and called for such defensive measures to render Soviet missiles obsolete[22].

If we analyse these twenty odd years of space history we find that there are not many civilian programmes which could benefit by the knowledge gained through the satellites except for getting weather information and certain information about the earth's resources. Every effort had been made by the super powers to vie with each other and get the strategic advantage. Countries with limited space fairing capabilities were left well behind. The policy of Mutually Assured Destruction (MAD) had given way to a newer policy of strategic defense as a means of offense. It may be mentioned here again that ABM treaty, signed in 1972 and revised in 1974, prohibited both the US and the Soviet Union to maintain more than one missile defence site and 100 interceptors each. Thus both were equally open to missile attack by the other side. And it left US vulnerable to Chinese attack with its 20 odd ICBMs. Following the Reagan pronouncements and fearing the Soviet reaction in May 1986, a large number of scientists and engineers including few Nobel laureates from universities around the US refused to accept SDI funds.

However, continuing with the militarisation of space, NASA and US Space Command joined hands and decided to:

- Launch Defense Support Programme satellites from the space shuttle.
- Use the Shuttle for Air Force technology payloads

NASA and the Ballistic Missile Defense Organization (BMDO) jointly sponsored the " The Deep Space Program Science Experiment. Statement[23] of General Joseph Ashy (Ex Commander in Chief US Space Command) makes the intention of US militarisation of space quite clear. He stated, "Its politically sensitive, but it's going to happen.... Some people don't want to hear this, and sure isn't in vogue, but absolutely we're going to fight in space. We are going to fight from space and we're going to fight into space. That is why the U.S has development programme in directed energy and hit to kill mechanisms".

Following this revelation a USAF reported," in the next two decades new technologies will allow the fielding of space based weapons of devastating effectiveness to be used to deliver energy and mass as force projection in tactical and strategic conflict."

US Space Command's Vision[24] 2020 is perfectly in place with the space policy that the US has been pursuing for many years. Methodology to be adopted involves the following operational concepts:

- Control of space, assured freedom to operate, deny the enemy.
- Global Engagement includes worldwide situational awareness, defence against ballistic and cruise missiles and the capability to hold at risk from space a small number of high value targets. Provision of dominant battle space awareness enabling on demand targeting and engagement of all ballistic and cruise missiles.
- Full force integration: War fighters to take full advantage of space capabilities as an integral part of special joint and combined warfare.

Nations following such developments made them fairly worried which was voiced by Kofi Annan in March 1999, at the UN' Conference on Disarmament. In November 1999 the UN General Assembly was asked to reaffirm the Outer Space Treaty[25] and specifically its provision that space shall be used for peaceful purposes and the " exploration and use of outer space shall be carried out for the benefit and in the interest of all countries. 138 nations voted for the motion titled " Prevention of an Arms Race in Outer Space". The US and Israel abstained.

Nuclearisation of Space[26]

Presently there are three nuclear devices used in space:

Radioisotope Heating Units: Plutonium 238 is used for heating the instruments in satellites for deep space probe

Radioisotope Thermoelectric Generators: Plutonium 238 is used to generate electricity to power instruments on board space probes. 48 lbs of plutonium used on the Galileo mission to Jupiter. 72 lbs were used for Cassini mission to Saturn.

Space Nuclear Reactors: These are used to generate large amount of electricity.

Accidents involving Nuclear Devices in Space[27]

The USA launched 24 devices carrying nuclear material and three have failed so far. The Russians have launched 39 and six have failed. In 1964 a satellite carrying plutonium powered system crashed on Earth with 2.1 pounds of plutonium on board. Apollo 13 in 1970 had 8.3 pounds of plutonium on board. It was ejected before re-entry in case the spacecraft burned up on its re-entry in the atmosphere and landed in the deep Tong Trench in the South Pacific. In 1978 a Soviet Cosmos satellite with a nuclear reactor crashed into the Northwest Territories of Canada. And Russian Mars probe crashed into Chile and Bolivia in 1996 with half pound of plutonium on board. Therefore, mankind should embrace itself for any catastrophe by these seemingly innocent objects circling overhead. As the plutonium is considered the most dangerous radioactive substance and if dispersed in an accident could affect a large population. NASA's own assessment was that, if Cassini would have broken up and the plutonium got dispersed, approximately 5 billion of the world population would have received 99 percent or more of the radiation exposure. On 28 January, 1986, seven astronauts and $100 million tracking satellite blew up 70 seconds after launch. And if this tragic incident would have happened to Challenger's next mission in May, which was carrying Ulysses plutonium, fueled space probe with about 24 pounds of plutonium onboard, the result could have been somewhat different. Even if we take NASA's statement at its face value, then we should be prepared for such nuclear accidents unknown previously. If plutonium rained down on areas of natural vegetation, it would be necessary to relocate animals immediately.

If it fell on urban areas, the requirement would be to demolish some or all structures and relocate population permanently. The next NASA space nuclear mission is the Europa Orbiter to be launched in 2003[28].

Ballistic Missile Defence (BMD)

BMD of USA consists of National Missile Defence (NMD) and Theatre Missile Defence (TMD)[29]. The defence is for the homeland or for the US fighting forces fighting away from home. The operational principle of NMD is that the Space Based Infrared System (SBIRS) will be made up of satellites in high altitude and low altitude orbits. SBIRS (high) will consist of four satellites in geosynchronous Earth orbit and two satellites in highly elliptical orbit. The low altitude SBIRS platforms will employ two sensors—acquisition and tracking—operating in a variety of wavebands including short, medium and long wave infrared. Launch detection will be the responsibility of the high altitude SBIRS providing the earliest trajectory estimate to command and control system. SBIRS low would provide mid course tracking and discrimination capability. Once the re-entry vehicle separates from the missile, the X band Radar would act as the primary fire control radar guiding the interceptor to the target[30]. Receiving cues from SBIRS low and the early warning radars, it would employ a narrow radar beam that can detect a typical warhead at a range of 4,000 km and is likely able to detect reduced signature target at 2,000 km away. At smaller distances it is said to be capable of discriminating between decoy and warheads. Ground based Interceptors will be launched at this stage, which is nothing but a land based missile intended to approach an incoming warhead outside the Earth's atmosphere and release the Exoatmospheric Kill Vehicle (EKV), which will steer itself, and impact with the target. EKV will receive mid-flight update from ground-based radars and satellite sensors. The plan is to base 250 interceptors both at Alaska and at Grand Forks Air Force Base in North Dakota. EKV is causing lot of concern to US developers[31]. The first EKV test failure occurred on 18 January,

2000, when a cooling line malfunctioned. Second one malfunctioned on 7 July, 2000.

Theater Missile Defence

Currently the US forces are using the upgraded versions of the Patriot and Hawk land mobile surface to air missiles. New design concepts are based on two tiered architecture consisting of upper-tier system that attempt to intercept an incoming missile either above or just within the earth's atmosphere and lower-tier systems engage the missile at much closer range. Army's upper tier system will consist of the much talked about Theater High Altitude Area Defence (THAAD)[32] system. Horizontal and vertical range of these hypersonic missiles is 200 km and150 km respectively. James Hackett reporting for *Washington Times* mentioned that on "10 June 1999[33] early in the morning Army's Theater High altitude Area Defense Missile interceptor streaked into the New Mexico sky and seconds later smashed into a modified Minuteman missile on the upper edges of the atmosphere more than 50 miles above the desert. The precision shown by one missile hitting another with two closing at a speed approaching 15,000 feet a second is mind boggling. The success of THAAD came after six straight failures and the project was under pressure to be closed down. The interception technique is the classic example of hit-to-kill technology, which requires the interceptor to find the target missile, tracks it at high speed, homes on to it, and finally destroying it by impact. In March 1999, Patriot exactly achieved the same."

The ideal scenario would be to intercept an incoming missile above the earth's atmosphere and in case of a miss, engagement of the same could be in the atmosphere itself but much above the surface of the earth. The second tier system working in conjunction with THAAD would be the Patriot advanced Capability-3 (PAC-3)[34] . These are the upgraded version of the PAC-2 used in the Gulf war. Post-Gulf war improvements made the PAC-3 to be the missile with hit to kill technology rather than its older version, which had a proximity fuse. This enables

PAC-3 to use smaller warheads and smaller boosters developed in the 1980s for the Strategic Defence Initiative (SDI) programme. The smaller size of the missile will have number added advantages operationally.

It may be pertinent to mention here that the Stinger, the shoulder fired missile, had 18 straight failures in its initial tests yet eventually perfected to become the weapon that contributed more than any other systems to the Soviet defeat in Afghanistan. THAAD is very important for the US forces because it will give them a upper level of defence that can reach out and intercept missiles much higher and farther away than Patriot. THAAD could provide air defence cover to a whole theater of operations, a large city or even a small country. While Patriot defends smaller high priority targets, a single THAAD battery could defend all of Taiwan and three batteries could give Japan the necessary Air Defence cover.

Laser in Space

The USA has been working on a laser based anti-missile system. Recently the Tactical High Energy Laser, developed in co-operation with Israel, shot down a missile in a test carried out on 6 June, 2000[35]. The system uses a deuterium fluoride chemical laser to shoot down medium to short range missile at a range of up to 5 km. The USA is also in the process of employing High-Energy chemical oxygen iodine laser mounted on a modified 747-400F aircraft with a capability to shoot down theatre ballistic missiles in their boost phase while remaining in friendly airspace. If all goes well with the development plan, then a fleet of seven Airborne Laser will be operational by the year 2008 with capability to shoot down missiles at their boost phase[36].

Russia in Space

As revealed presently, the Russian space programme was never been the result of centralised planning that many US space analysts feared. Every USSR space achievement had been sending ominous chill through the spine of the Western military planners. First Sputnik, first man into space, the Soyuzes Vostoks etc., kept pace with the US achievements. Even ten years ago, the awesome Russian annual launch rate was sustained at well over 90 orbital flights per year, picking at 101 in 1992[37]. It has since fallen to less than half, which could not keep all the factories in business. In the year 1999-2000, the Russian launch rate had fallen below the level in the yearly 1960s when race for the space had just begun. Post-Soviet revelations have brought certain startling facts. Throughout the best days of the Space age, Moscow managed its space programmes within ad hoc link up of agencies, funding authorities and review boards. Long term Soviet space doctrine was set more by whim and misperception rather than rational consideration. Organisations like Glavkosmos (international sales office) were created to deal with public or even international partners. But neither they had the expertise nor were keen for expanding business, etc. Leadership chain was not very clearly defined, and bureaucratic mishandling was the order of the day. More so, Kremlin's whims and fancies created serious disruptions in the entire space industry. The fall of the Soviet Union has seen the change in the organizational efforts. Moscow has created the Russian Space Agency called RKA[38] (the initials for its Russian name) headed by Yuri Koptev. RKA at present controls the civilian space budget, both manned and unmanned. Military Space agency is known as VKS (initials for Russian name). VKS headed by Colonel General Vladimir Ivanov runs the nation's space tracking networks and has its own spacecraft control centre Gloytsino-2 near Moscow[39]. It also manages the Plesetsk Cosmodrome, Russia's northerly launch site, which had been under the administrative control of the Russian Strategic Rocket Forces. Civilian RKA and military VKS share the Baikonur facilities and RKA reimburses the VKS for the wages of more than half the military workers there for their logistic support for civilian launches. RKA and VKS jointly fund and manage the Gagarin Cosmonaut Training Centre on an air force base northeast of Moscow.

Launch Facilities

From 1957 through 1980 at the height of the cold war era, Moscow played a deception game by indicating one of its space centers at a different place while building the same near the railway station of Tyratam East of Aral Sea near 46 degrees North. To disguise the name, the authorities named the same as Baikonur, the name of a small village 250 kms to the northeast[40]. Even the official applications for international space flight records carried false latitude and longitude.

The second USSR rocket center was a converted intercontinental ballistic missile base at Plestsek[41] , above Moscow and to the South of Arctic Circle. When this center began to launch satellites with Cosmos in 1966 another deception game was played by the Soviet Union. The name was never published till 1983. Despite this unacknowledged Plesetsk launch site, it accounted for more than half of all space launches in the whole world with Baikonur coming second to it. Plesetsk concentrated on routine launches of medium/small boosters and simple payloads, while complex manned and deep space missions were launched form Baikonur. Advantage of Plesetsk is that it is in Russia and had been holding on its own despite staffing problem and other shortages like power cut offs due to unpaid bills, etc. Its winters are hard but Baikonur located further south suffers much more from harsher weather. But Plesetsk also has certain disadvantages. It is at latitude of 62 deg North. Its rockets cannot reach the planned 51-degree inclination of the International Space Station without orbital plane change, which consumes lot of fuel. High latitude of this cosmodrome makes it extremely costly to handle payloads aimed at high altitude geosynchronous orbit useful for communication. Moreover, Plesetsk is not yet equipped for launching big boosters. It can at the best launch medium sized Soyuz and Molniya boosters. Launching pad for bigger Zenit boosters is under construction. The completion time can be anybody's guess. Proton is the Russia's most powerful operational booster, which is about the size of Titan IV, or Saturn I. Launching of the same from Plesetsk is unlikely in view of the massive cost of building proper infrastructure. Therefore, only location to launch Protons will be Baikonur which is deep inside the now independent Kazakhstan. And if Baikonur also goes out of the hands of the Russians due to its ageing equipment, much of which is rapidly wearing out and breaking down, its decline in personnel whose living condition is at a most pitiable state in the workers' city of Leninsk[42], and finally if there is any political trouble with Kazakhstan then Russia with its Proton rockets will have no where to go to sustain its space programme. The current treaty allowing Russia to lease Baikonur cosmodrome from Kazakhstan was based on negotiations between the two prime ministers of these countries at an annual rent of $115 million based on the estimated interest of a Kazakh debt to Russia of more than $1 billion which Nazarev, the Prime Minister of Kazakhstan, acknowledged as part of his deal with Yeltsin for his country's independence.

Russian Space Boosters[43]

Proton has been the biggest and most reliable Russian space booster in operation. Built at the Khrunichev rocket factory in Moscow it now has more than 200 launches to its credit. To keep the facilities going, commercial deals have been signed with European and US communications satellite builders like Inmarsat, Huges, Astra, etc. Alongside this Proton upper stage, Lavochkin design bureau in Moscow is working on an independent alternate upper stage named Fregat, which can be used with all the Russian boosters not just the Proton. But unless the cash flow by marketing the same is ensured the Fregat take off would be doubtful for quite sometime to come. The Russian launching continuity could be attributed to two reasons. Firstly, few new rockets are rolled off the assembly line by a small but dedicated but poorly paid staff. And secondly, the exploitation of the strategic reserve of boosters that were stockpiled long ago for military support. The stockpile is supposed to be continuously replenished to retain strategic advantages. But that did not happen. And it is estimated that the stock has been reduced to zero recently, which delayed the manned mission to Mir space station.

Beside the Proton rocket built at Khrunichev, other booster factories are upgrading their products. Soyuz/Molniya booster is being upgraded to a new version called Rus because it is based on all Russian components. The intermediate Kosmos booster is undergoing similar changes to become the Vzlet space booster similar to the US Delta launch vehicle. Russian rocket builders hope to make profit from the direct sales of rocket engines to Western aerospace firms. However, sales of cryogenic upper stage engines to India created Washington-Moscow crisis. The US feared that India could convert these rockets into military missiles. And finally Russia succumbed under the US pressure and decided to withhold this manufacturing technology from India.

Manned Space Programme

Russian Air Force had been the host of Gagarin Cosmonaut Training Centre[44] in a small town near northeast of Moscow. The center is in complete disarray due to lack of maintenance. As no funding was made available to the Russian manned space programme, the officials learnt to offer guest cosmonaut flights at substantial fee to West European countries. $ 85 million dollars were earned by the center by providing two long flights to European Space Agency astronauts in 1994 and 1995. In 1993 NASA signed a $ 400 million contract for extended US visits to the Mir space station[45]. It seems that Russian manned space programme can probably survive on these cash flow conditions. In a recent development Russia decided to launch the US civilian millionaire 60 years old Dennis Tito to space station in a Soyuz TM-32 space vehicle for an undisclosed sum of money[46]. The cosmonaut corps endured a large layoffs of older men in their 40s or 50s, many of them never got chance to fly but were highly trained. Support for space medicine research was also drastically reduced, groups like Institute of Medical and Biological Problems, received no state funding at all.

Despite such serious setbacks, Russian military still maintains its own reconnaissance satellites designed to verify arms control treaties and keep an eye on the deployment of Western strategic weapons along with network of communications, missile warning, meteorology, and navigation systems, geodetic payloads and improving ICBM aiming[47]. And until recently, a special fleet of Russian satellites turned their sensors on Russia's own domestic targets to monitor military units for any lapses on countermeasures against the US satellites. Yet in the past five years, in addition to reduction of trained personnel, their budget has been slashed to a tenth of former levels in actual purchasing power. Russian space center Kaliningrad control center is the one shown off to the world but it is not the main space station of Russia. Although Kaliningrad controlled the space station, the other 160 active Russian satellites are controlled out of Golytsino. This Russian Military Space center is about 45 km to the west of Moscow.

Financially, the civilian programme is slightly better off than its military counterpart. Some programmes have been abandoned for lack of funds. Western funding has been invited for many programmes. Yet, the Global Navigation Satellite System (GLONASS)[48] has completed its operational constellation. It is estimated that GLONASS has 100-150 meter accuracy for non-military users. It is estimated that Russia had 24 operational geostationary spacecraft, 18 out of these are beyond their designed life[49]. The constellations are normally replenished at a rate of six to eight spacecraft per year but the replacement launches have been falling every year and as a result some serious gap will be created in Russian communication capabilities. Russian officials themselves forecast the collapse of many of the networks and the loss of services. At that point of time, they will either have to do without the services, or pay for the same to get the services from the Western neighbours. Mir itself is feeling the budgetary stranglehold. The loss of experienced personnel has affected the operation of this space station seriously. Not even the Russians know, how long the overworked and inexperienced ground controllers will keep the Mir functioning before some disaster overtakes it.

Old Soviet space programme was kept in total secrecy and was designed to mislead, either to allow protection of real technology or to trick the foreigners into overestimating the level of Soviet space technology. But as the communism collapsed so did the system of deception upon which it relied. Old memoirs and photographs flooded the post-Soviet news media. Tales of space disasters and debacles became a common story with the Western press. On the face of such financial crisis and fast decaying space programme and overaged satellites, it would be anyone's guess, how long the Russian space industry will be able to maintain their once super power status in relation to the rapidly advancing space militarisation by the USA. There are no indications whatsoever that the Russians have any response to the much publicised US NMD programme or offer any counter to space domination as articulated by Rumsfeld in his 8th May speech.

China in Space

Although Beijing has received some degree of foreign technological assistance in the areas of reconnaissance, surveillance and targeting capabilities many of its system development efforts appear to have a substantial indigenous component. China already has launched three low-orbit meteorological satellites and a geosynchronous weather satellite[50].

Bill Gertz reported in *The Washington Times* on 23 February, 1999, that " China is sharing space technology with North Korea, a move that could boost Pyongyang's long-range missile programme.....The reports are alarming because North Korea used a long range Taepodong missile in August to try to place satellite in orbit. They also heighten fears among some Pentagon officials that militarily useful U.S satellite technology shared improperly with China in 1995 and 1996 may have been given to North Korea[51] ."

No doubt, China will be a force to reckon with. But Chinese dream to challenge the USA in outer space may not be realized for many years. China has not been a large contributor to the militarisation of space so far. But no doubt it has the capability to launch military photoreconnaissance satellites but these may not have much consequences on the face of present days developments in space. Beijing could eventually deploy advanced imagery reconnaissance and earth resource systems with military applications. The China-Brazil Earth Resources Satellite (CBERS)[52] was launched in October 1999, and the experience gained from operating this satellite will support Beijing's efforts to develop improved military reconnaissance satellites. China also may attempt to deploy a near-real-time electro-optical imaging satellite within the next decade. Till such time, China will have to make good with resources from commercial satellites like SPOT and Landsat imagery.

China is interested in 400-500 kilogram (kg) satellites and plans an oceanographic research satellite Haiyang 1 (Ocean-1) scheduled for launch this year. Beijing eventually may launch satellites for earth observation, communications, and navigation. China also is developing mini-satellites (weighing less than 100 kgs) for missions, which include remote sensing and networks of electro-optical and radar satellites with 50 mtrs resolution. Satellites in this series probably will have improved resolution at a later date. Beijing also participates in the Asia-Pacific Small Multi-Mission Satellite Project as part of the Asia-Pacific Multilateral Cooperation in Space Technology and Applications Program, which reportedly includes Iran, Pakistan, Thailand, Mongolia, South Korea and Bangladesh[53].

China is yet to have the capability to conduct "launch-on-demand space launch operations," i.e., the capability to use satellites and space launch vehicles in storage to launch within 24 hours of a decision to do so. All major aspects of China's manned space program began within the last five years. A manned space launch may be expected any time during this year (2002) or early next year. This has nothing to do with space militarisation. This would be to demonstrate technological advancements. China's manned space efforts could contribute to improved military space systems in the 2010-2020 time frame.

Beijing may have acquired high-energy laser equipment and technical assistance, which probably could be used in the development of ground-based ASAT weapons that could destroy satellites in the future. Specific Chinese programmes for laser ASAT have not been identified. Only the press articles indicate an interest in developing this capability. *Washington Times* confirmed the assessment when they reported on 3 November, 1998[54], " China's People's Liberation Army is building lasers to destroy satellites and already has beam weapons capable of damaging sensors on space based reconnaissance and intelligence systems. Consequently China could blind U.S intelligence and military space equipment systems vital for deploying military forces in current and future wars."[55]. Beijing already may have acquired technical assistance which could be applied to the development of laser radars used to track and image satellites and may be seeking an advanced radar system with the capability to track satellites in low earth orbit. It also may be developing jammers, which could be used against Global Positioning System (GPS) receivers. But it is also known that PLA is using US GPS and GLOSNASS satellites in developing advanced weapons and may use these satellites to improve accuracy of its missiles.[56].

China's aerospace industry is seeking to integrate GPS and Russian Global Navigation Satellite System (GLONASS) guidance technology into fighters and helicopters. The China Aerospace Corporation displayed a GPS receiver at an exhibition in Beijing in September 1996 and provided brochures advertising both a 12-channel GPS receiver and a 12-channel GPS/GLONASS receiver.

To broaden technological base for space, China has been working assiduously and joining in all forms of joint ventures with Western countries. Since 1998, China and the United States have signed three inter-governmental agreements on satellite launching services. Several US satellite-manufacturing companies have signed agreements on commercial satellite launching services, involving a total of some 30 satellites. The satellite "Sinosat"—jointly developed by China, France, and Germany—was launched successfully in 1999. Moreover, in the form of a joint venture,

China and Germany have made improvements to the Dongfanghong 3 communications satellite and have worked on a new generation of similar satellites[57].

An Analysis

Our study of the militarisation of the space and its repsercussion may have been well established by Uday Bhaskar in his article "Bush Outlines New US Nuclear Policy" in *Strategic Analysis* wherein he states " ... it is pertinent to note the salience being accorded to the medium of space in the new US security policy. Assets in space such as satellites and related surveillance capabilities will be the key to battlefield domination in the future and the Bush administration is moving with determination in this regard. On May 8, a week after the Bush speech, US Defence Secretary Donald Rumsfeld announced a series of organizational changes to sharpen the Pentagon's focus on managing defence interests in outer space[58] ". Complete military domination of the outer space has been the clearly defined long-term goal of the USA. From the early 60s the USA has been gradually building this capability while keeping an eye on the development by the other states. The US Space command's vision 2020 is no doubt another pointer to this respect. The USSR was a serious counter to all the US plans, and by various means, the Soviets kept the United States guessing about their technological advancement. But since the post 1991 development, the Russian race for space has been lost forever. Cash strapped Russia will never be able to challenge the US superiority in space. No Outer Space Treaty will be able to check this space domination. China, a serious contender for the next super power status, may have developed few ballistic missiles to threaten the USA but the space capability of China is nowhere close to that of the USA. Yet the USA will not be in a position to discount China totally because of the reported development of Chinese capabilities in laser technology and jamming capability of the satellites. Indian space programme has been a story of success but till recently Indian military did not have any participation in ISRO programmes. Services have started taking interest to take the space factor in their doctrines which is rather late to contain the US designs. Indian military will be lucky to find a parking slot at the geostationary belt, which is getting crowded every day. Satellite signal interference will be the order of the day unless the military establishments keep themselves abreast of the current developments and take active steps immediately.

However, complete domination of outer space by the USA will not be as simple. Russia still has some ASAT capabilities, which was articulated by Boris Yeltsin in 1998, but he went on to report that Russia had abandoned the tests. If this statement is read carefully and knowing Russian solid base in field of Physical science, such reports cannot be discounted immediately. A completely beleaguered Russia may devote its scientific brain to develop laser and particle beam weapons to strengthen anti-satellite systems. China is believed to have already progressed a lot towards the development of ASAT laser weapon as reported in various the US newspapers. If the report has even 25 percent accuracy then the dream of Rumsfeld may not come true for a long time to come. 21st century may brace itself for another kind of warfare not known to mankind before.

Notes:

1 "Defence Secretary Rumsfeld Outlines Space Initiative" Official Text, Public Affairs Office of US Embassy New Delhi, May 9,2001.

2 "Space History 1 – The Evolution of Space Power" (*Federation of American Scientists*, February 2000)

3 Ibid

4 n1, p 2.

5 Dave Webb, *The Nuclearisation and Militarisation of Space- A Brief History.* (Leeds, England)

6 Ibid p3.

7 Ibid p7.

8 Garwin RL and Bethe HA, " Anti Ballistic Missile Systems" *Scientific American*, March 1998.

9 Ibid

10 n2, p7.

11 Ibid

12 n1, p2.

13 n5, p8.

14 Ibid

15 Hilsman Roger, *Cuban Missile Crisis: The Struggle Over Policy*, Westport, Praeger pub, 1996.

16 Outer Space Treaty. The Treaty on principles governing the activities of states in the exploration and use of space including the Moon and other celestial bodies came into force on January 27,1967. The substance of arms control provisions is in Article <*h.ttp://www.state.gov/www/global/arms/treaties/space1,htm*>

17 Ibid

18 n8.

19 Daryl G Kimball and Stephen W Young," National Missile Defense and Arm Control After Clinton's NMD Decision" Geneva, United Nations Disarmament Forum One-*2000.*

20 Jeffery R. Barnett, *Future War,* Alabama, Air University Press, Maxwell Air Force Base, January 1996. pp123-125.

21 n5, p11.

22 Ibid

23 n5,p16.

24 US Space Command vision 2020 www.spacecom.af.mil/usspace/visbook.pdf.> Also, Karl Grossman writing for Knight Rider/Tribune news mentioned in November 22nd issue that "Vision for 2020 compares the US effort to control space with the effort centuries ago when the nations built navies to protect and enhance their commercial interests by ruling the oceans.....Just look at the material coming out of the Pentagon's US Space Command....cover of the Vision for 2020 report of the Space Command for example depicts a laser weapon shooting a beam from space zapping a target below. The report opens " US Space Command dominating the Space dimension of military operations to protect US interests and investments. Integrating Space Forces into war-fighting capabilities across the full spectrum of conflict.".<*http://www.globenet.free-online.co.uk/articles/arace.htm* >

25 Karl Grossman, "Arms Race in Outer Space" Knight Rider/Tribune News Service 22 November22,1999.<*http://www.globenet.free-online.co.uk/articles/arace.htm* > The report stated that " On November 1 the United Nations considered a resolution entitled prevention of an Arms race in Outer Space." The resolution which 138 nations voted for, said that space " shall be used for peaceful purposes". The United States was not one of these nations, however, it cast a lonely abstention. This was a shameful vote, and it puts the United States on a path toward weaponization space. Actually, the United States is on that path already."

26 n5, p 17-19.

27 Ibid p 21-22.

28 n5, p19.

29 George N Lewis "US National Missile Defênce Option" *Pugwash Occasional Papers* Vol.2, Number2 March 2001, pp15-31.
30 Ibid
31 ibid
32 John Pike and Peter Voth " Current Plans for Missile Defence" Geneva, United Nations Disarmament Forum One –2000, pp1-5

33 James Hackett, " Theatre High Altitude Area Defense" *The Washington Times,* June 10, 1999.
34 Ibid
35 Marie Hewish "Beam Weapon's Revolution" *Jane's International Defense Review, August 2000.*
36 Ibid
37 James Oberg, "Russia's Space Program: Running on Empty- Part 1" p7 < www.astronautix.com>.
38 Ibid p2.
39 Ibid
40 Ibid pp4-5.
41 Ibid
42 Ibid pp4-5.
43 N37, p6.
44 James Oberg, Russia's Space Program: Running on Empty – Part 2 < www. astronautix.com>.
45 Ibid
46 News from Russia, Vol IV No 17 April 27,2001.
47 Ibid
48 Ibid
49 Ibid
50 Annual Report on the Military Power of the People's Republic of China, Persuant to the FY 2000 National Defense Authorization Act (Section 1202)
51 Bill Gertz, *The Washington Times,* February 23, 1999.
52 n49.
53 n49.
54 *The Washington Times,* November 3, 1998.
55 Ibid
56 Annual Report , Government of India Department of Space 1995- 1996.
57 Annual Report , Government of India Department of Space 1997-1998.
58 C Uday Bhaskar, " Bush Outlines New US Nuclear Policy, *Strategic Analysis, Vol XXV* June,2001.

10

Air Power Of 21ST Century:

An Indian Perspective

"Air power today has reached the status of the dominant factor bordering on a pervasive impact on war. It is true that air power may not win a war by itself alone; but without it no major war can be won[1]."

"Geographers and sailors, had long been aware that three quarters of the earth's surface was covered by sea; it took a little while for the realisation to sink in that all of it was covered by sky. No longer would a river, a hill, a forest, a sea or ultimately even an ocean impose an arbitrary obstacle to the conduct of warfare[2]."

" A 43-day air campaign brought about the withdrawal of Iraqi troops from Kuwait[3]."

"Air power searches enemy, not only on the battlefield but right inside the safety and security of his homeland. It is capable of hitting him where it hurts the most[4]."

"The history of success of air power and the importance of air power have now become enshrined in the doctrines of war fighting[5]."

The war strategists' thought that Douhet was over enthusiastic when he proclaimed that air power could be decisive in all future wars. Some even ridiculed him. On the contrary, results of World War II proved the contributory factor of the air power.

The might of air power was displayed fairly conclusively. Histori pre-emptive strike by the Israeli Air Force on June 5, 1967, an application of air power by the Indian Air Force under th leadership of Air Chief Marshal PC Lal are the glaring example of application of air superiority doctrine. Finally, Gulf war prove that Douhet was correct. Application of air power could b decisive.

Iraqi military and political leadership was totally surprise by the successive waves of air attacks made by hundreds c aircraft and Tomahawk missiles, which destroyed Iraqi comman and control centres, Scud missile sites, air defence networl airfields and aircraft on the ground.

What do we understand by air power? Does the air powe refer only to combat potential of an air force based merely o fighting element and the combined air arm comprising the fighter transports and helicopters available with all the three service Or is it something else? As per our assessment air power is a all pervasive factor consisting of its different entities bearing o its effectiveness, employment doctrine, training imperative operational concepts, tactics, leadership and practical experience The combined effect of these elements could provide the desire result from the employment of air power. And such relate factors are mostly common to all the air forces of the world

Till 1971 Indo-Pak war the terminology used was th "number of sorties per kill," but since 1980s this has taken different connotation when it is termed as "kill per sortie," effecte by enhanced battlefield information, real time intelligence an application of precision guided munitions. Such capabilities wer well established during both the Gulf and Kosovo wars. Nigh attacks by B2 bombers and F117 stealth fighters displayed thei mettle by attacking the enemy targets effectively. With th introduction of Precision Guide Munitions (PGM), battlefield visio elements have become extremely important to provide timel and accurate information about the enemy military assets. Ai campaigners of this century will hanker for real time informatio and its processed formats collected through surveillance reconnaissance, signal intelligence, aircraft radar navigatio equipment transmission and communication device intercepts. The list is virtually endless. There is no doubt that real-time intelligence will be the pre-condition for effective employment of the air power at any stage of the war.

It is also understood that air warfare is an activity in which all services have equal and important part to play. It is not air force centric. Air power entails harnessing every related component, inclusive of space and information warfare elements. It may be interesting to note here that during operation Desert Storm, it was not a F117 which delivered the first Laser Guided Bomb (LGB) but a US Army Apache helicopter launched the first attack with a Hell fire missile on a Iraqi forward air defense post[6].

Despite Israelis routing the Arab air forces in June 1967 by surprise attacks, air power had never been so eulogized as it happened post operation Desert Storm. Making war in the third dimension had taken a quantum leap. Allied air campaign against Iraq demonstrated for the first time a breakthrough in the strategic offensive effectiveness of air delivered weapons. Bombing offensive[7] against Germany during World War II had been the precursor of the strategic offensive of the air power, which was nothing but serial operation like any other military operations. When majority of the enemy systems lay in relative safety and most of the conflict was restricted to the front only. Despite its success in Second World War, such strategic effectiveness got a heavy beating during three years of operation "Rolling Thunder,"[8] against North Vietnam. All such lessons had to be reexamined when the Iraqi forces were completely routed and allied established complete air control over Iraq followed by the eviction of Iraqi troops from Kuwait. A *coupe de grace*, effected by the air campaign. This was termed as the final coming of age of air power. Billy Mitchel[9] and Douhet could prove these concepts to a large extent, when Douhet commented in his book *The Command of the Air*,

> "... the purpose of aerial warfare is the conquest of the command of air. Having gained the command of the air, aerial forces should direct their offensive against surface

targets with the intention of crushing the material and moral resistance of the enemy.[10]"

Serial attacks in Germany gave way to parallel warfare, which brings many sections of the enemy system under simultaneous attack. It renders the system inoperative and cannot react correctly to defend itself. Even repairing of the system or part thereof becomes extremely difficult. Best example of such parallel war has been the strategic attack on Iraq in operation Desert Storm. Within a short time the coalition air forces attacked over a hundred key targets, comprising Iraq's entire strategic depth. Following the attack, Iraqi air defense centers faced problems on detecting or guiding any aircraft, telephone services were dislocated, lights went out, key leadership and personnel were isolated. But during the entire cold war period, strategic air warfare was only related to intercontinental range bombers and nuclear weapons. These were considered to be an essential section of nuclear deterrence. All other elements of air power other than these strategic component formed the tactical air power or in the US parlance " Theatre air power." The entire effort was to be utilized towards supporting armoured or infantry formations in their land battles in the form of Close Air Support (CAS) or at the most in Battlefield Air Interdiction (BAI). Little was known or thought about the employment of air power beyond such concepts. Classical example may be the employment of Indian Air Force in 1971, which had only the tactical component in it.

The air of change began to blow, sometime during the early 80s. When it was realized that air weapons are getting more lethal with improved accuracy. These have imparted to conventional air power a qualitative improvement in its ability to achieve better results. This was confirmed by the allied air campaign in the Gulf war, dispelling many doubts that grew out of Vietnam about the use of air power. Air Vice Marshal (Retd.) RA Mason summed it up by stating, "the Gulf war marked the apotheosis of 20th century air power[11]."

As it stands to date, strategic attack and the fragility of the states at the strategic level had been established in post-1991 era. If examined closely it may be seen that the countries are balanced most precariously which stands on their strategic bases like leadership, communication, key production, infrastructure and population. And if a country loses its ability to defend itself against air attacks it would be paralysed and wouldn't be in a position to sustain its armed forces on the battlefield. That is why the first task for any government would be to protect the citizens and critical infrastructures. Once a country loses strategic air superiority it would be at the mercy of the enemy. Therefore, while on the offensive, the commander should try to attain strategic air superiority in contrast to gaining tactical favourable air situation, as has been the principle till 1991.

These focus on the fact of another new terminology "Parallel Warfare". The idea may be new but proved conclusive in the recently concluded air wars. Strategic targets of each country are not many in numbers and they wouldn't have many aiming points for destruction or being crippled. These targets are not very large in size, are very expensive, do not have much redundancy and are difficult to repair. Once these targets are engaged in a simultaneous manner or struck in parallel the damage to the nation could be overwhelming. In comparison, if we take the case of attacks on Germany during Second World War, the serial attacks were carried out on one or two targets allowing enemy to disperse or strengthen defenses of similar targets elsewhere in the country. That is why serial attacks do not produce the desired strategic result. Enemy can concentrate his resources to repair damage to a target already under attack and conduct counter offensives whereas, parallel attacks deprive him of such ability. From here the effectiveness of Precision Weapons come into focus and the theory of "kill per sortie" rather than "sorties per kill." Nature of war has changed from probability to certainty. Strategic, tactical or economic targets difficult to conceal will become victim of these precision weapons, fielded since mid-eighties. In all the wars known to mankind, each side fired very large quantities of projectiles with the hope that enough damage will be caused for the enemy to retreat or sue for peace. War was in the realm of unpredictability rather than the predictable design as of today guided by the Precision Guided Munitions provided these are delivered correctly at the correct targets. Economy is the main principle of employment of air power in war and Jasjit Singh points out rightly that,

> "There is also the question of another time tested principle of war, that of economy. Nowhere does it apply with greater force than in relation to air power[12]."

Air power doctrine essentially includes principles of war and the unique capabilities of air power like, flexibility and versatility, concentration of force, centralized control and decentralised execution. However, the concept of air superiority occupied the central theme of all air operations.

Following Iraq's invasion of Kuwait on August 2, 1990, General Horner while briefing on the air operations plan to President Bush emphasised on gaining air superiority[13], interdicting the attacking Iraqi forces and defending the ports and rear areas. He also presented a second plan of offensive retaliatory strikes against targets in Iraq, if Iraq used chemical weapons against the US forces. As a matter of fact the General remained well within the classical air power doctrine as discussed earlier. But shortly after this invasion, a group of staff officers in the Pentagon began planning an air campaign designed to eject Iraqi forces out of Kuwait. Col. John A Warden III, the Air Staff's Dy. Director for War fighting, supervised this effort. The group initially had only six officers but quickly grew to more than one hundred officers from all the services; this ad- hoc group became known as Checkmate[14] named after the facility of an Air Staff Division. This group rationalized a concept of operations and a targeting scheme to accomplish the final objective using air power all by itself. The plan developed by Warden and Checkmate was named "Instant Thunder[15]," which called for an intense six-day air campaign designed to incapacitate Iraqi leadership and destroy its key military facilities. Their principle was to attack the centres of gravity rather than the fielded forces, the destruction of which would force the enemy to yield to US wishes. Gen Schwarzkof called the VCAS of Air Force Gen John M Loh for help in planning air operations who in turn subscribed to the ideas of **nstant Thunder**. Warden model[16] is explained below:

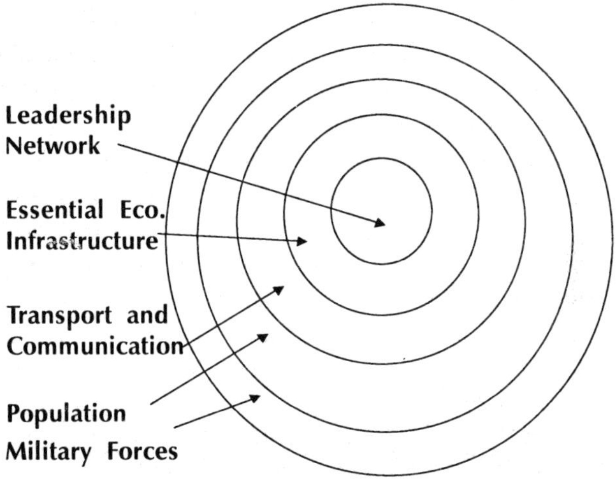

Innermost circle in Warden model is the leadership. They considered Iraq as a system and the entire system interdependent. They ignored the individual targeting but fixed the target points only derived from the system. These points if targeted will cripple the system. And those were the presumptions. It gives us the idea of the relative importance of each entity contained within a given ring. Essential economic structures if destroyed could reduce the enemy's war waging potential but may not have any immediate effect as expected by the air planners. They also surmised that the objective is to convince enemy and make him to do whatever they wanted him to do. The entity of the power to agree to change is the leader in the center. Thus the entire energy of the war should be directed to change the mind of the leader. The rings clearly show that the military is a shield for the whole system and not the basics of the system. Therefore attacking the tanks and aircraft may not yield the final result of bending the will of the enemy. The rings give us the concept of working from the inside to the outside as opposed to the general concept of attacking the fielded military first before getting to the leadership. Warden made this concept essentially strategic rather than a tactical approach. Therefore, for using the concept it would be essential to start with the largest identifiable system. That is why when Iraq invaded Kuwait, it was identified as a system rather than the Iraqi forces in Kuwait. Five-ring analysis gave the planners a complete picture about the targets to be attacked. Such type of analysis could be continued till the level of an individual mission flown by a four aircraft or two aircraft. What the Checkmate actually said is that the five-ring system were to be applied to every system even if you want to attack an electrical system. Several five rings may have to be designed to get to the lower subsystems. The method may be differentiation and not integration. Entire principle is based on the centres of gravity, which is nothing but organizational concept. Which ones are most important becomes clear when we decide what effect we want to produce on the enemy in order to induce him to accept own position. And finally what to attack becomes a matter of the capacity. Warden concept may be listed as under considering that the five rings will be good method for identifying centres of gravity:

- The enemy is a system and not made of tanks and aircraft alone. Military is the protecting shield around it. Engagement of the military will have its effect but not directly.
- Enemy leadership is the principal opponent to your cause and should be punished to bring them around to see your viewpoint.
- Five-ring concept may be applied to any system as such to identify the centers of gravity and to identify the targets to attack.

Warden briefed AF Hqs and Joint Staff and insisted that **Instant Thunder** would force Iraq to withdraw from Kuwait. The planners believed that it would be necessary to attack Baghdad and Iraq proper and not solely the forces deployed in Kuwait. The plan had 84 targets to strike over six days. **Instant Thunder** offered the US high command the only offensive option but, it was initially rejected by General Horner who was the Commander of the Central Command at that time. But once he discussed the plan with the other air commanders, he subscribed to it totally because he himself didn't have an alternative answer. The air campaign of the Desert Storm planners consisted of the following strategic targets[17]:

- Command and Control and Communications.
- Leadership facilities.
- Nuclear, Chemical and Biological warfare capabilities.
- Military support facilities.
- Ballistic missile launchers and their supporting infrastructure.
- Electric power.
- Oil refineries.
- Key bridges and railway facilities.

Attack against the leadership and command control had political and military dimensions. Separating the national leadership from the forces in the field would delay the coordination of military operations and show the Iraqi forces the powerlessness of their leadership. To prevent propaganda, the attacks were targeted against radio, television transmitters, relay stations, telephone and telegraph facilities and military command posts. Planners hoped that the air campaign would cause the removal of Saddam Hussein. Planning of the second, third and fourth phases of the air campaign, which dealt with Iraqi forces, were also considered unnecessary by the Instant Thunder planners.

These eight targets contained 295 target points initially on 15 January, 1991, and 535 on 26 February, 1991, due to continued intelligence update throughout the 43-day war. These eight targets absorbed about 15 percent of the coalition air to ground efforts[18].

The planning of the second, third and fourth phases of air campaign dealt with the Iraqi forces in Kuwait. Air superiority became the aim of Phase II only at the suggestion of Gen Schwarzkof[19]. The General went in for the classical application of air power and didn't want any interference in his ground offensive by enemy air force. Phase III consisted of battlefield preparation which tasked air power to destroy ground forces to a degree not planned by any air force before. Battlefiled Air Interdiction (BAI) is again a time tested theory of application of air power. Coalition air power was given the unenviable task in Phase III to destroy 5,000 pieces of dug-in and defended Iraqi equipment. If all went well the final phase ground offensive operation could commence as and when the situation permitted.

Desert Storm and its Aftermath

1991-Gulf war has now come to be viewed by the analysts as having achieved much less[20] than it was expected to do. The assessment has brought out that the coalition attack on Iraqi leadership, telecommunication, command and control was not an unqualified success as was expected[21]. At the end of the war, Saddam Hussein frustrated all US effort to dislodge him. On the other hand he came out much stronger than before in the eyes of his countrymen. People of Iraq somehow understood him more than the USA did. There was no uprising to overthrow him. Saddam got away by blaming the USA for all the hardship it caused to Iraqi people.

Despite the continued US interference, Iraqis could continue to launch Scuds even during the final days of the campaign. Communication was no doubt disrupted between Baghdad and its field army in Kuwait theater of operation but enough redundancy must have been there for the Baghdad leadership to order withdrawal and redeployment. These results bring direct criticism of Warden's application of air power theory. The ambitious theory did not fulfill the results. It in no way proved that bombing the leadership, command and control system will bring about overthrow of the regime and completely sever communications between the leadership and their military. Iraqi national telecommunication systems supported by enough redundancy could recover quicker than was estimated by the coalition military planners. Fibre optic network and computerized switching system systems proved difficult to put out of action.

Attacks on Electricity and Oil[22]: Coalition forces mounted about 890 strikes against electric power and oil installations. About eighty percent of Iraq's installed capacity was damaged or destroyed and the national grid was isolated. But to the surprise of the entire world Iraqi grit to restore commercial power proved to be successful and the estimated period for recovery of Iraqi electrical facilities proved to be entirely out of place. By mid-1992, electricity was restored almost to about 90 percent of the pre war capacity. This proved beyond doubt that air power inflicted little damage to the Iraqi power system, and could be restored without external assistance.

Similar was the case with the oil production. It was estimated that 90 percent of Iraq's petroleum refining capacity was made inoperative. But barely within two years and by October 1992 Iraqi officials claimed that the production had returned to 800,000 barrels a day[23]. The refineries began to supply enough petrol and other products for domestic consumption and for exports. The skeptics put it that, since the Iraqi Air Force was on the ground as such there was not much of requirement of fuel. Such type of assessment puts the assessment characteristics on dubious footing. The attack on oil installation might have been devastating for the time being but the military planners did not easily understand recuperation period. Militarily this could have been effective only if the war lasted longer than it happened.

Attacks on Chemical Weapon Targets and Scud Sites [24]: Stockpile of Iraq's chemical weapons came out in the open after the war. It was so concealed that the entire stock of the weapons remained untouched, which they did not use against the coalition forces. The attacks against the chemical warfare facilities disrupted the production facilities but could not entirely stop it. Similarly, the efforts to suppress Iraqi Scud launch sites were not a complete success. As a matter of fact, coalition forces' efforts to find and destroy the mobile launchers proved elusive and frustrating. The launch rate per day declined over the course of the Gulf war but the actual destruction of any Iraqi mobile launchers by fixed wing aircraft remained impossible to confirm. The fundamental sensor limitations of coalition aircraft, coupled with the effectiveness of Iraqi employment tactics (including the use of decoys), reflects that a few mobile Scud launchers were actually destroyed but the number fell well short of the claimed kill by the pilots.

Attacks on Ground Forces[25]: Attacks on ground forces consisted all the classical components of application of air power like Battlefield Air Interdiction (BAI), Close Support to Army and Navy, etc. The vast weight of the air effort got directed against the Iraqi ground forces in Kuwait and against the supply line. Republican guards were singled out as the center of gravity This portion of the air war was characterised by the gradual attrition of the Iraqi forces deployed on ground. Battle damage assessment also took the route of assessing the overall damage caused rather than sortiewise details. The bombing altitude and slant range was increased to avoid attrition of allied aircraft, which affected the effectiveness and accuracy of the weapons.

Air strikes against Iraqi ground forces began on the very first day of the air campaign and continued with increasing intensity throughout the war. The air planners aimed to reduce Iraqi ground potential by 50 percent. Statements by Iraqi prisoners of war indicated that the sense of futility dawned on the ground troops after weeks of extensive bombing and they appreciated the fact that their equipments were the target rather than the personnel themselves.. The true effects of these attacks came not from specific targets destroyed but from the combination of targets attacked and the intensity with which the attacks took place.

Close Air Support(CAS): CAS system during the Gulf campaign followed a classical pattern when the attacking forces enjoy a total supremacy of the air. Such a thing couldn't have been possible even in Vietnam despite overwhelming supremacy of the US forces as they didn't enjoy the complete air supremacy.

Formations would arrive at a particular contact point from the target area and loiter for the commander to ask for an attack on a particular target; if no targets were allotted the fighters would carry out search and strike missions and engage targets of opportunity or carry out attacks under the control of the Fighter Air Controller (FAC).

Lessons of Desert Storm: The air operations in the Gulf war have come to be viewed as having been achieved considerably less than a towering policy success as promoted by Col. Warden and Checkmate[26]. Overenthusiastic lessons and the accolade for air power have to be examined very carefully before drawing any final conclusion for any other operation. The conflict embraced an unusual combination of circumstances, which might not be repeated elsewhere. This was proved during the Kosovo air campaign. The opposing air forces were badly mismatched; asymmetry of forces gave the coalition an upper hand before even the war began. The Iraqi leadership failed to appreciate that the country well suited for ground offensive also can be the victim of ground attack aircraft. This was a clear case of suicide when the Iraqi forces moved forward and then dug in and waited for the allied forces to build up their strength constantly to their ultimate disadvantage. Saddam Hussein declared, " The air force has never been the decisive factor in a battle in the history of warfare[27]." His limited knowledge of air power had been a source of strength to the Iranians who fought Iraqi aggression for eight long years. Saddam Hussein blundered again. That is why, application of air power in Desert Storm was for limited objectives and the lessons were anything but conclusive. The main cause of the Iraqi Army's breakdown was not the destruction of their tanks and artillery pieces, as the individual soldiers' desertion. The rate of deseration of Iraqi troops reached an alarming proportion due to relentless air attacks on their facilities and supply lines.

War of Kosovo

Principles that were derived following the Gulf war were again put to test during the Kosovo war. The air power's recently acquired fame almost clearly came to undo the entire concept. But finally air power prevailed. From March 24 to June 7, 1999, NATO, led by the United States, conducted another air campaign without any ground deployment. Air power succeeded again. But in-depth analysis of the application brings out many flaws. The first week only saw about 50 sorties a day in sharp contrast to Desert Storm, which mounted more than 800 sorties a day. From these 50, half returned without delivering their loads due to bad weather. Though the air effort was increased subsequently, casualties on the ground became limited due to successful Serbian dispersal tactics and more robust air defenses than were mounted by Saddam's forces. On top of this, there was no ground threat to Serbian forces. As a result they could survive the air attacks by dispersing and concealing their armour and other vehicles[28]. The missed attacks on a train carrying passengers and the Chinese embassy by the much vaunted Laser Guided Bombs (LGB) became an embarrassment of the entire campaign. The target system consisted of oil refineries, petroleum depots, road and rail bridges over the Danube, military communication sites and factories capable of producing weapons and spare parts. But Milosevic's strategy was one of playing for time and frustrating NATO. He almost succeeded till such time the NATO decided to attack the power generation capability. Towards the final stage of the war, the Yugoslav electrical grid was put out of action completely on May 24, which affected Yugoslavia badly. The entire banking system came to a grinding halt along with other electrically operated important national systems.

Suppression of Enemy Air Defence (SEAD) faced many problems and the campaign did not go as planned like in Desert Storm. The Serbs kept their SAMs hidden with their radars not emitting. Probably they had been waiting for the NATO forces to commit to lower altitudes where they could be more easily engaged. But it was also not clear whether the SAM units could not operate due to their disruption of communication linkages.

The problem became acute, as the Joint STARS who carried out a tremendous job in Desert Storm could not locate the target due to mountainous terrain covered with vegetation. It allowed the Serbs to lie low and set traps for the allied fighters and then shoot and move out to another location.

Nothing was working out for the NATO forces in comparison to the Gulf war. The five ring concept of Warden was taking a heavy beating. Air power was used as the situation demanded. Indian Air Force will have to learn from these experiences and should not depend entirely on the Checkmate planning of Gulf war. The efforts to attack the communication links of Yugoslavia's Integrated Air Defence System (IADS) was not disrupted because of latter's underground command sites, buried land line and mobile communication centers. They also applied the tactics of fused radar input. The radars located in the North were carrying out the tracking and acquisition feeding the same to the radars in the South which enabled the Southern operations centres to train their defensive weapons without having to activate any of their radars, thus avoiding the attacks from Homing Anti Radiation Missiles (HARMs). These may be the reasons why the F-16s were less effective with their HARMs to home on to enemy early warning radars, which normally operated close to the SAM batteries. Jamming effort by NATO was also countered by only activating the fire control radars, which made them susceptible to HARM attack. But activation of Serbian fire control radars was sporadic as such the HARM shots had to be reactive rather than preplanned leading to many failures. Finally, NATO fighters had to go in for the hard kill of the enemy defensive radars. Whenever enemy radar transmission was picked up the fighters were directed to attack the site, provided they could reach the ground point in time. On further analysis it was brought out that the allied fighters orbited at a far away place from the SAM sites and by the time they could be activated and reach the target, it was too late to carry out any effective attack. Whenever the SAMs engaged NATO fighters, they had to defend themselves without any assistance from the airborne command and control.

This effective air defence tactics finally bore result when a F 117 was shot down by a barrage of SA-3s. This shooting down of a nearly invisible aircraft by antiquated Soviet SAM system created much concern in the allies' camp.

Serbian determination and application of operational tactics destroyed many air power theories, which came out of such air defence capabilities as displayed in the Kosovo conflict. The allied forces were forced to climb to higher altitude to seek the enemy defences and other ground targets. It was not easy for the NATO fighters to detect and carry out the attacks effectively in mountainous terrain while flying at 15,000 feet which raised the basic question of adequacy of the Suppression of Enemy Air Defence (SEAD) tactics being employed. These also brought out the necessity of real time intelligence about mobile radars and other assets and inform the pilots quickly for them to react. It took many days to locate Serbian air defence emitters and disseminate the intelligence to aircrew, which was in sharp contrast to operation Desert Storm experiences where the Iraqi air defences almost gave themselves away to the attacking forces whenever they resorted to transmission of the radars.

Effectiveness of Attack on Ground Forces: Initial NATO claim of damage caused to the enemy tanks and other vehicles was mostly unfounded because of the successful dispersal and concealment of tanks and guns. Camouflage technique limited the capacity to carry out air attacks against fielded forces in the absence of any associated friendly ground threat. They could hide their armour in relative safety which otherwise they would have had to deploy to cut off the roads and bridges to the advancing NATO troops on the ground and then make themselves vulnerable to air attacks. Instead, the Serbs hid their tanks, used smaller vehicles to go into villages to terrorise them. Famous "Tank plinking" did not work at all in the entire Kosovo war as it did in Kuwait war. The Serbs also had skillfully employed decoys in large numbers showing targets like tanks and SAM sites, deflating instantly when hit by NATO fighters. That may be the result of large differences between the pilot claim and the actual destruction of ground targets. It proved beyond any doubt that a determined and skillful enemy could survive air attacks. Once again flying at 15,000 feet the NATO pilots could not determine the type of vehicles civilian or military carrying the refugees. The entire campaign was opportunity based rather than on any well-formulated plans as claimed following operation Desert Storm.

> "..senior US military officials say they have identified several critical problem areas that emerged from the bombing of Yugoslavia[29] ."

It was exactly after 78 days of bombing of Serbian military, the sudden decision of Milosevic to give into the NATO demand came as a reprieve for the beleaguered allied air power. He agreed to an internal peace plan and withdrawal of his forces out of Kosovo.

What could have been the turning point? The military analysts place this on the Kosovo Liberation Army (KLA) offensive on May 26. This thrust was met by the Serbs with their tanks and artillery thus making them vulnerable to enemy air attacks, despite making sure that they move in small groups to avoid detection. UAVs and forward air controllers did a fine job to detect and direct firepower onto Serbian ground forces, which could not remain in hiding any more. This proved that even after 70 days of air attacks, Serbs had enough amount of hardware to fight back. Yet, Milosevic realized the futility of the ongoing attrition and decided to give in.

The allied forces flew a total of 37,465 sorties out of which about 14,000 sorties were flown for strike purposes, 23,000 munitions were expended, out of which, 3000 to 5000 were delivered against 500 odd fixed targets[30] .

In such a sudden turn of events the air power again came out unscathed as the super hero of the campaign. The US newspapers who doubted the campaign from its very beginning showered praises about the achievement of the air power. Someone called this as the century's most impressive air campaign. *The Wall Street Journal* stated that it is " one of the biggest victories ever for air power[31] " But there were doubts about the efficacy of a campaign with air power alone. As the war progressed it was proving frustrating for the NATO forces because there was no viable ground threat to Yugoslav forces. The employment of forces was definitely flawed though the result was favourable. The attacks on the Serbian forces were ineffective and the atrocities against the Kosovar Albanians increased as the air attacks intensified. Air power can certainly deliver punishing attacks but cannot change the mind of the leadership. On the other hand targeting choice made by the NATO commanders did not follow any pattern or support or any given theory. Warden theory of Desert Storm did not have any chance in Kosovo air war and does not point towards any future application of air power. It could have failed miserably because defensive operation had not been catered to by Warden and his compatriots. They only theorized on Saddam Hussein's lack of knowledge of air power and its capabilities. Meager achievement of the first month of the NATO forces convinced Milosevic that he could survive the NATO air offensive and increased his ethnic cleansing. Probably the NATO offensive failed to see the importance of surprise and the axiom of keeping the enemy guessing about its intentions.

In the absence of any ground threat Milosevic could have wethered the NATO attacks by digging in and carrying out his ethnic cleansing which prompted NATO forces to carry on with attacks and retain its cohesion. However, one can very well point out that but for the air power Milosevic would have carried on with his atrocities is true but air campaign's targeting had not received universal approval, and mostly by the air power analysts. And if anyone comes to a final conclusion that it was air power alone which won the war, it would be very simplistic and narrow a view. Many contemporary factors along with the air campaign helped to make his final decision to surrender. Russian disapproval, international outcry and UN declaring Milosevic a war criminal were the factors that weighed heavily against him. NATO ground intervention threat was also becoming a reality with troops being organized at various levels, The *coup de grace* was the surprise KLA ground offensive during the last two weeks of the air campaign which drove Serbian troops out of hiding and made them lucrative targets for the allied airmen. It seems that an offensive build up on the ground could have made the Serbian forces more targetable. This would have also kept them away from carrying out crimes on the civilians as they found time earlier to do with impunity. This typically brings out the effective application of both ground and air in synergy rather than application of air power in isolation. Air power cannot perform its best unless a credible ground component is available for the campaign. The incremental strategy chosen by the allied forces could have been a disaster followed by the spectacular success in Desert Storm. Therefore it should be understood that " military power in the modern world is useful only in two circumstances: in self defence or as a way of conquering another country. It will be useless if the aim is only arm twisting[32] ."

Air War in Afghanistan

Wars are not unknown to Afghanistan. Afghan people have been fighting incessantly for a number of years and avoided being subjugated by any power in recent history. The Soviet Union with its massive show of strength had to finally withdraw from Afghanistan. But the recent war waged by the USA against the Taliban regime had been unique. The war had been waged by air power and special forces. Carrier based aircraft had conducted the major offensive due to absence of suitable land bases in the region for tactical strike aircraft. Heavy bombers like the B1 and the B 52s had played critical role, dropping overwhelming quantity of bombs on the designated targets. These giant aircraft have effectively been used for close air support missions, aided by the special forces on the ground to designate the targets.

Unmanned Aerial Vehicles (UAV) like Predators had played an extraordinary role which for the first time delivered Hellfire missiles against the ground targets followed by the Global Hawk with its wide angle of vision used in any combat zone.

Enhancement of computational power of the computers and networking and broad band satellite communication assisted intelligence from the UAVs and other sources to be relayed to the command posts. Multi-source information processing has to go a long way but it was effectively tried out in the Afghan war. And it is said that the critical sensor to shooter time period had been reduced considerably. A relatively minor role was played by the land based F16s and F15s. The US carrier based aircraft mostly took part in the operation but at a massive cost. On top of these a massive number of tanker sorties were needed to get these aircraft to the target and recover them to their bases.

However, it does not need an air power pundit to point out that Afghanistan's primitive air defence did not expose the US warplanes to any kind of risk that they would face elsewhere. It would be grave mistake if it assumed that other countries will also be poorly defended in any future operation.

And finally it should be understood that it was not the air power that fought alone in this war. There was a ground war fought by Northern forces and there was significant deployment of US special forces.

Lessons from Gulf, Kosovo and Afghan Wars

Applicability of Warden Model for Indian Air Force: Warden model of five circles should be analysed carefully before it can be adopted in the Indian environment. Concept of operation will centre on massing of fire rather than forces. Strategic air superiority vs tactical air superiority is the key question to be analysed and answered. The Checkmate group depended heavily on the offensive concept leading to strategic air superiority in preference to any defensive plan whereas Indian Air Force will have to cater for both offensive and defensive operations and mainly depend on tactical superiority which will result in dilution of available limited forces. Despite the versatility of air power, certain amount of defensive elements will have to be committed to safeguard the Indian sky and vital installations. IAF will never be able to muster such preponderance of forces like the coalition forces in operation Desert Storm. Achieving strategic air superiority over Pakistan or China will be an impossible task.

Warden's war objective had been to convince the enemy leadership that it may not be worthwhile to continue with the belligerence. In the case of Indo-Pak or Indo-China conflict, it would be very difficult to analyse correctly the system and the organisation as a whole. Some guesses could be made only if we try to analyse the situation from the IAF point of view. There is no doubt that the survivability of any system depends on the survivability of the leadership and other elements of the five rings. But how to proceed with the concept? Such a concept becomes a victim of its own when the IAF segregates its forces for defensive purposes as well. The shield has to be fielded otherwise the system could come under severe strain from being attacked by the enemy. In addition, IAF do not and will never have preponderance of such force level to give shape to this strategic concept. It could only be possible for the USAF and US Navy combine and on the face of general asymmetry of forces as prevailed during Gulf war. But the concept is no doubt very clear and may be that many subsystems which constitute war efforts directly or indirectly could be analysed and center of gravity determined to get to the specific target points. This was the major reason for acceptance of the idea by veterans like Gen Honor prior to the war. But the strategic application of the same left many doubts in the mind of the air power theorists. Saddam Hussein is very much alive and came out much stronger than before. The problem with Saddam Hussein was that he never understood the air power. He relied heavily on his ground and missile forces. But IAF cannot be placed in a similar category. It understands the importance of the application of air power but is badly restricted in its operating capability due to many factors like:

- Too much to defend.
- Too much to attack.
- Too little a force.
- Too less money to augment force level.
- Too difficult to engage the Command and Control at every level.

Yet, five-ring concept could be applied to an extent for identifying the centres of gravity for each target system.

Indian Air Force will have to understand that air power is becoming more and more victim to its well-earned name and fame. The consistent accuracy of the weapon system, little collateral damage will become the norm of every operation led by the airpower. Nothing less will be acceptable to the international community. Collateral damage caused during Desert Storm and recently concluded Afghan war caused much embarrassment for the allied air power. Many sorties were aborted in the Kosovo war for fear of collateral damage when the target could not be identified. This could not have been a good example following the grand success in Iraq. The statistics of flying more than 30,000 sorties and losing only two aircraft during the entire Kosovo campaign shouldn't be taken as an example of force protection. It makes to look as if air power is indeed risk free. Similar example may be quoted from the Afghan experience of the air power. While minimising one's loss is admirable it should not be forgotten that the forces were never equally matched. Indian Air Force will have to take all these factors into consideration and enhance its operational potential and do not become a political victim of a situation beyond its control.

" By now historical standard we've been able to restrict the scope of violence to the people who have it coming to them, in an extraordinary way," said Elliot Cohen, a professor at John Hopkins University's School of Advanced International Studies.[33] There is no doubt that the bombs have become smarter. Precision guided munitions have gotten more precise and flexible. In the Gulf war it was estimated that ten percent of munitions were PGMs. In war of Kosovo, the majority of US bombs were laser guided variety, which ran into trouble due to bad weather. On the contrary in Afghanistan the bombs were all weather and satellite guided. The Joint Direct Attack Munitions (JDAM) a $ 20,000 guidance kit, transformed the ordinary iron bombs into an accurate munition using satellite linked global positioning system to guide it to the target, though it was less accurate than the laser guided bombs[34]. The advantages of the same are that these are much cheaper and can be dropped by a wider range of aircraft. But it should not be forgotten that these are not bug proof. Weak batteries can break the connection between the satellite and the bomb. Human errors can result in feeding wrong coordinates.

The next most important lesson of this war had been the improvements in target information gathering and processing. In the USA this has been termed as "network centric warfare[35]." As such, the demands on reconnaissance and surveillance planes like Rivet Joint, JSTARS and AWACS would increase.

Bombers like B1s and B52s played a very high profile role and were used for close air support missions, whereas fighters planes, the heroes of the Gulf war, had much lesser role to play because of their shorter ranges and the long distances for the Afghan missions. As a result mid air refueling was in great demand which helped some of the bombers to operate from the continental USA. Also, a warning for the fighter pilots had been sounded in this war when the Predators (UAV) were used effectively to deliver Hellfire antitank missiles controlled by someone at a computer on the ground.

But it is again cautioned that overplay of the dominance of air power in preference to all other forces would be incorrect. The Marine Gen. Peter Pace, Vice Chairman of the Joint Chiefs of Staff, said "Sometimes the best weapon system is an individual with a rifle, other times the best weapon is a plane with a bomb on it[36]."

Stephens Alan of Australia has commented that "..modern war is concerned more with acceptable political outcomes than with seizing and holding ground[37]." Whatever may be the lessons or indictments on air power, certain points can speak for themselves very clearly when we examine the effort and the result quotient of the last century's air power application vis a vis the contemporary wars of the late twentieth century. Earlier air campaigns were of limited effectiveness because the aim was to vanquish the enemy on the ground facing own troops and try to punish the enemy by diverting certain air efforts to attack few strategic targets. The effects of these attacks were not very clearly understood. It took too many aircraft and too many losses to gain some seeming advantages. On the contrary, the air power of post Gulf war can make its presence felt much quicker and can influence the course of the battle sufficiently to derive the political aim of the nation.

Massing of force has been the greatest factor of the principles of war. But the capabilities of the present generation air power has proved beyond any doubt that massing of force can be carried out effectively with the deployment of accurate weapons systems. These could cause more damage by the lesser number than a large bomber forces deployed during the Second World War. Increased battlefield awareness, better aircraft survivability, and precision guided munitions are the major contributors towards this current generation understanding of massing of effort rather than fielding many fighters or bombers with a hope to cause certain amount of damage and making sure that some of the force actually survive to deliver some of the munitions they carried. The target can be actually identified, its vulnerability analysed and weapons employed to precisely decapacitate the same. Force application will naturally flow out of the same. Naturally, it would be much less than the force level of yesteryears. Air power had the supporting role to ground battle rather than any of its independent application. It has unburdened the ground commander to launch a frontal attack and accept grievous loss to human lives.

Achieving air superiority is a part of air power theory today. It is necessary but may not be sufficient for air power to lay convincing claim to having become the only force. Moreover, such employment of air power for inflicting pain by attacking economic or other symbolic targets considered vital to the interest of the enemy will be insufficient because air power could carry out much more damage throughout enemy's battle space to help outcome of the battle and affect the enemy directly. Punishment theory of the Second World War is no longer valid.

Stealth, precision guided munitions, battlefield awareness all help to apply air power in a totally different form, than was envisioned by the early proponents of air power. But in no way it suggests that air power can win a war without ground and naval involvement. Also it does not suggest that in all engagements, air power will be more important than the other two components of military power. That is why an aerospace based defence strategy for all occasions will be fraught with certain danger. Benjamin Lambeth stated that,

> ".... Air power now has the ability to carry out the bulk of responsibility for beating down an enemy's fielded forces of all kinds, thus enabling other force elements to achieve their goals with a minimum of pain to gain, effort and cost. More than that, one can argue that air power in its broadest sense, including such vital adjuncts as surveillance and reconnaissance in addition to combat platforms, munitions and mobility assets needed to deploy them, has fundamentally altered the way major wars will be fought over the next decades through its ability to carry out functions traditionally performed at greater cost and risk by other force elements. Most notable among these are its demonstrated capacity to neutralize an enemy's army with a minimum of casualties on both sides and its ability to establish the preconditions for achieving strategic goals from the very outset of fighting. Thanks to these new capabilities, air and space power, coupled with information power, now offers the promise of being the swing factor in an ever-widening variety of circumstances. .....It does not challenge the valid point made by surface warriors that the object of future land warfare should be to make close-in-killing a coup de grace rather than a bloody battle of attrition[38]."

Asymmetry of force level in both the wars, vis a vis the air power ascendancy will become a nightmare for the Indian Air Force. It will never command such a preponderance of force level as was the case, yet it would be expected for her to live up to the reputation enjoyed by air power.

Notes:

1. Air Commodore Jasjit Singh, *Air Power in Modern Warfare*. 1985. Lancer International presentation, New Delhi. p21.
2. AVM RA Masson, *War in the Third Dimension, Essays in Contemporary Air Power*. 1986.Brasseys Defence Publishers, London, New York. p2.
3. Air Cmde NB Singh, *Air Power in the New Millennium, 2000*. Manas Publications, New Delhi p171.
4. SN Rampal, *Air War*. 1974. New Delhi, Trimurti Publications Private Limited, p5.
5. n1, p21
6. Benjamin S Lambeth, Understanding the New Face of Modern Air Power, RAND, Santa Monica California, Paper presented on 15-16 November 1999 at Rydges Canberra Hotel, Canberra, p12.
7. n3. P 43. " The contribution of air power to the Allied victory during the second world war confirmed the views held by the early air power advocates."
8. AVM Marshal JE Johnson, *The Story of Air Fighting*, 1975 Hutchinson, London, pp231-246.
9. General Billy Mitchell was a military pilot with great experience of air operations. He was the pioneer to develop tactical air warfare doctrines.
10. Gulio Duhet, *The Command of the Air*, New York, 1942, pp102-103.
11. AVM RA Mason, RAF (Retd) "The Air War in the Gulf," (*Survival*, May/ June 1991) p225.
12. n1, p32.
13. Thomas A Keany and Eliot A. Cohen, Gulf War Air Power Survey Summary Report. 1994. Air University USAF, Maxwell Air Force Base, Alabama, p33.
14. Ibid p36.
15. Ibid
16. Ibid pp37-39.
17. Ibid
18. Ibid pp63-65.
19. Ibid pp47-52.
20. Benjamin S Lambeth, Understanding the New Face of Modern Air Power, RAND, Santa Monica California, Paper presented on 15-16 November, 1999 at Rydges Canberra Hotel, Canberra, p14.
21. Air Cmde N.B Singh (Retd) *Air Power in the New Millennium,2000,* Manas Publications, New Delhi 189. The analyst claimed that " Gulf air campaign shows that it achieved total success." Which is contrary to the analysis given by us. Air power did not achieve total success and it had its failures as discussed in the chapter.
22. Thomas A Keany and Eliot A. Cohen, Gulf War Air Power Survey Summary Report. 1994. Air University USAF, Maxwell Air Force Base, Alabama71-74.
23. Ibid
24. Ibid pp79-83.
25. Ibid pp91-98.
26. n6, p14.

27 John Godden, Shield and Strom London, Brasseys 1994, pp12-13.

28 Eric Schmitt, Bombs are Smart, but People are Smarter, July 4, 1999 at http://www.nytimes.com/library/review.

29 David A Fulghum and Robert Well, Weapons, Intelligence targeted in Probe, *Aviation Week & Space Technology* July 26, 1999.

30 Air Cmde N.B Singh (Retd) *Air Power in the New Millennium, 2000,* Manas Publications, New Delhi.

31 Thomas E Ricks and Anne Swardson, " Military, Diplomatic Offensives Bring About Accord'" *Wall Street Journal,* June 4,1999.

32 Thomas Abraham, The Lessons of the Kosovo War, *Hindu,* New Delhi, May 17, 1999.

33 Nancy Benac,War in Afghanistan Demonstrates Air Power's New Ability, The Associated Press, USA, December 19,2001.

34 Ibid p2.

35 Ibid.

36 Ibid p3.

37 Alan Stephens, *Kosovo, Or the Future of War,* Paper no 77, Air Power Studies Centre, RAAF Fairburn, Australia, August 1999 p21.

38 n 20, p29.

11

Challenges Of Air Land Battles

Despite having an enviable tradition of democratic governance since independence, four wars were thrust upon India. Lessons from these wars and the wars of the late 20th century require a complete analysis. Gulf war, Kosovo war and Afghan war have brought out many facets of the technology driven network centric war. Indian armed forces would have to understand these wars and implement the concepts like Revolution in Military Affairs (RMA) to engage in any future conflict in the subcontinent. Doctrinal issues are to be reassessed. It is fairly clear that the days of serial warfare is over, substituted by much talked about parallel war or hyper war. Therefore challenges for the Indian land forces and the Air Force are immense, accentuated by the scarcity of the resources made available for the armed forces. But that should not be a stumbling block to face the challenges of the coming years.

Challenges of Land Battles

A 1.1 million strong army consuming more than 60 percent of the defence budget has been on constant alert since Indian independence. It has been responding to every conceivable situation facing the nation both from within and without. Lord Wavell, when he made his farewell speech, stated, " I believe that the stability of the Indian Army may perhaps be the deciding factor in the future of India[1]." Probably he understood India more than the political masters of independent India. If an honest assessment were made it would be found that it is the Indian army on whose shoulder the Indian national aspirations had been riding safely. The chaotic democracy had been a success story only because there is a solid organization which never asked a question to its political masters and sacrificed their lives whether in North East, Sri Lanka, Kashmir or to help the civil administration These men in olive green uniform never shied away form its responsibility to come to the aid of civil power and bears the distinctive mark of respect among the comity of nations for its commendable job for carrying out UN peacekeeping in every conceivable part of the globe.

Tehelka like manipulators tried their level best to bring schism among the ranks of the army without realizing that this is an institution which never sleeps to guard an almost a 11,000km long boundary passing through the most inhospitable terrains in the whole world. Without realizing the effect it would have on the men fighting a relentless battle on the world's highest battleground at Siachen, where a normal human being will not be able to survive for more than few minutes without assistance.

While India shares and borders with seven neighbours, but few of them share borders among themselves, which brings to focus our special relationship with all these neighbours, (3244 km land border with Pakistan, a 4056km difficult and disputed border with China and a 4351km border with Bangladesh). Asymmetry of size, population and economy in relation to these neighbors also creates many problems, which do not have any readymade solution. The opposition political parties of these neighbouring countries always criticize treaties with these countries. Inferences of Indian regional hegemony are resented by most of them and Pakistan clearly defies the theory totally. It has not forgotten the dismemberment of East Pakistan

Security Environment

India faces several internal challenges as it attempts to progress and seeks to gain a rightful place in regional affairs. Poverty and population are primarily responsible for slowing down the country's modest growth added with political instability and absence of a strong central government. Illiteracy also remains another major hurdle for social upliftment and checking population growth. Cross-border terrorism, external interference, ongoing insurgencies in North Eastern part of the country, and Sri Lankan ethnic clashes spearheaded by Tamils of that island will pose many security related problems to this country. India is an energy deficient country and over dependence of oil import will have many security ramifications. And while commenting on the stability in South East Asia, Ashley Tellis commented that,

> " South Asia will experience ugly stability for an extended period. The ugly stability will be characterised with the under mentioned:
>
> (i) Premeditated conventional conflicts of both unlimited and limited aims will remain absent.
>
> (ii) Security competition will continue through sub-conventional violence.
>
> (iii) Both India and Pakistan will remain incapable of attaining what may be the desired political objectives through war.
>
> (iv) Nuclear capabilities in the region have proved more destabilizing than ensuring stability. This has enabled Pakistan to support insurgency within India to settle outstanding political differences and wear down India. And prevent India from a conventional strike for fear of a nuclear holocaust.[2] "

Indian Army has the primary responsibility to guard a land frontier of over 11,000km. This includes a live and hostile border with Pakistan, which extends over 3200km and the difficult border with China, which extends over 4351km.

With both India and Pakistan becoming overt nuclear powers, there has been a subtle change in the military threat scenario. In any future Indo-Pak conflict, resolution of the J&K problem, degradation of Pakistan's military potential and military defeat would be of prime concern. These three are not mutually exclusive. Absolute decisive victory under the current geo-strategic and regional environment may not be feasible. On the conventional front, therefore, our ability to launch punishing punitive action to force Pakistan to abandon her support to 'proxy war' in Kashmir and elsewhere in India is of prime importance.

In the context of land-air operations, this would translate into securing of permanent gains in J&K and sensitive strategic areas in Punjab and Sindh, besides achieving a major operational level victory in desert sectors. Assistance in achieving and maintaining the initiative, leading to destruction of Pak military infrastructure and forces and prevention of inter-sectoral response of Pak forces by the Indian Air Force and blockade of Pak sea lanes to cripple her logistic staying power by the Navy would form part of overall objectives.

Thus, besides the need for urgent review of our war doctrines to take into account technological and regional politico-military developments, we need to redefine our strategic objectives in order to prioritise funding weapon development programmes and modernisation of the army.

In addition, the army of India has constantly been engaged in internal peacekeeping and counter-insurgency operations since independence. As a matter of fact Indian army may be the most experienced in the whole world on both these counts. Indian army has been contributing to UN peacekeeping most willingly whenever called upon to do so. Therefore, it may stated that:

- Indian military had been in the midst of Low Intensity Conflict (LIC) since Indian independence. Starting with Naga insurgency, Indian military is now facing an uphill task to contain cross-border terrorism in Kashmir valley. Despite accepting heavy losses on manpower and equipment the armed forces are supporting civil authority to counter LIC. And unless and until a definite political will is displayed, the Indian army will continue to fight this low intensity conflict for a long time to come.

- India has always been giving support to the causes of United Nations and as indicated earlier it would be in the best of interest for India to continue to do so.
- As Huntington stated, civil authorities will always bank upon armed forces to help them during national emergencies like cyclone, flood or earthquake, because of their organizational strength.

Immediate Challenges

Technological advancements in the recent years have had a profound impact on the battlefield and the milieu is going to be more and more sophisticated in the next two decades. Network centric war is the order of the day as the lessons learnt from the recent conflicts along with the induction of cutting edge technologies in the field of surveillance, targeting, avionics, weapon lethality and weapon delivery. To meet such diverse requirements the army will have to develop the following:

- The present surveillance capabilities of the Indian army are not sufficiently strong enough to guard thousands of kilometers of border, which runs mostly along difficult terrain. Unless effective surveillance capabilities are created at the earliest, Kargil episode could reoccur to the much embarrassment of the national leadership.
- Army has to assess whether it has the capability to fight non-linear battle in future, in order to engage the enemy both at intermediate and depth echelons simultaneously. If not, then it has to build up the capability to fight the wars of this century effectively.
- Needless to state that, to achieve quick responses, army should acquire greater mobility and greater firepower in both offensive and defensive formations.

- Enhance mobility in the plains and in the desert so as to exploit opportunities effectively.
- Good amphibious capability to counter any threat to our Island territories.
- Besides modernisation and induction of force multipliers, there would be need for streamlining, restructuring and reforming the army with a view to absorb future advanced technology platforms and systems and to implement adjustments that will be required by a shift in doctrine for war. More focus on air and sea control operations is inevitable – which in a sense would reflect a renunciation of our long standing ambition to marginalise Pak war making machinery and occupation of maximum territory in J&K and elsewhere. Notwithstanding the above, as long as conventional conflict remains a possibility, the army will have to maintain capability to conduct 'punitive' actions to force Pakistan to abandon support to proxy war in Kashmir and elsewhere in India.

Pak Military Responses

The assessed Pak military aims would be to:

- Safeguard its sovereignty and defend the territorial integrity of Pakistan, including Pakistan Occupied Kashmir and Northern Areas.
- Continue its ongoing 'proxy war' in Kashmir as its preferred option and attempt to spread its impact in other areas (including North East) to tie down regular forces of the Indian army in large numbers.
- In the event of conventional war being forced on her, adopt overall reactive strategy and attempt to end the war in the shortest possible time, by using the veiled threat of nuclear weapons and/or achieve a stalemate. Extension of the war beyond Pakistan's logistic stamina has to be avoided at all cost. In case of an imminent

Indian attack, pre-emptive operations would be a distinct probability.

In pursuance of her overall military aims, Pakistan's perspective plans would be to maintain conventional forces at levels which enable her to remain unmindful of our retaliatory capability, thereby enabling her to continue her 'proxy war' and raise its ante with relative impunity. To achieve the above, she will attempt to:

- Raise a third strategic reserve (Army Reserve Sindh).
- Raise two additional Infantry divisions for its existing Army Reserve North and South.
- Raise two to three additional Armoured Brigades for its holding formations. Make up existing shortfalls in combat support formations/units.
- Modernise her tanks and APC fleets.
- Pakistan has no apparent definition of nuclear threshold. There is however every likelihood that she will not resort to using nuclear capability, due to serious international fallouts, unless a strategic collapse is imminent or her heartland is directly threatened.

Changing Nature of War

It would be difficult to identify political reasons to carry out any protracted war with our belligerent neighbours. And also war may not be taken to the logical conclusion of military victories as was displayed during Kargil conflict. The world community will definitely step into to limit the scope. Therefore the war in this century will be conducted to meet the political objectives in preference to military victory.

It may also be seen that warfare has moved down the paradigm scale of intensity and inclusiveness. Potential nuclear warfare has given way to restricted nuclear deterrence. Use of military power for coercive diplomacy has increased, without going to war actually.

Limiting factors like avoiding a nuclear exchange will influence any future war in the subcontinent. The problems of escalating cost of weapon systems and maintenance of the equipment will have their own limiting dynamics.

On top of it all, battlefield environment will go through a sea change due to induction of new technology. Long range precision and lethal capability combined with a wide range of delivery systems is emerging as key factor in future wars. Multi-spectral sensing, improvements in information and computational power etc., integration of technologies, other adaptations to take the forms of increased mobility, manoeuvre war in all types of terrain will hold the key. Future wars will be characterised by greater dispersion, stealth and higher tempo of operation with less need to mass forces. Delays in induction of futuristic technologies will make the Army obsolescent and would seriously affect its operational capabilities.

> "..the speed, range, precision and devastation of modern weapons system are rapidly changing the nature of war and making the traditional Indian strategic posture more difficult. The Gulf war showed clearly the enormous power of offensive weapons. Advanced armour, precision guided missiles, long range missiles and modern aircraft will require Delhi to revise its defensive concepts[3] ."

Force Levels

Types of Forces Required: The battlefield of the future would have a much greater transparency, where an effectively integrated command and control system would give instantaneous analysis and immediate options. Integrated battle groups with state-of-the-art equipment and missile support systems would be required to deploy and redeploy quickly. The battle would be intense with sudden escalation of situations, relentless operations by day and night, and massive destruction during engagements. Integrated forces, with the required air and ground mobility, as also a quick build-up capability, would be needed to be able to exploit the opportunities presented. Additional VA's could come up in our EEZ and the importance of Island territories would increase accordingly. The force structure of the Army would be required to have the following:

- An increased surveillance capability, both in the hills and the plains sector, and correspondingly a larger number of EW units.
- The ability to conduct a non-linear battle, so that enemy's intermediate echelons can be tackled simultaneously.
- Greater mobility and hitting power in both offensive and defensive formations in order to have quick response and maximum destruction capability.
- Intra-theatre mobility at formation level in the mountains as well as in the plains sector, so as to exploit opportunity or restore an adverse situation.
- An amphibious capability to defend our Island territories.

Shortcomings in Present Structure: While there is no requirement for recommending any major accretions in force levels or manpower requirements, yet due to the resource crunch, there have been slippages in modernisation and conversion. The major inadequacies are as follows:

- There is no capability to carry out non-linear battle in the absence of an airmobile formation, and Air Force resources do not permit the strategic lift capability for reserve formations.
- There is no offensive capability against China.
- There are serious voids in our existing field formations.
- The command and control capability has been curtailed and both the mountain and the infantry divisions have an inadequate surveillance capability.
- Adequate amphibious capability does not exist and this affects the defence of our Island territories.

- Other modernisation aspects such as introduction of Smart ammunition, providing desired mobility to our ground forces, development of EW potential and force multipliers, have been slowed down due to the lack of funding.
- Infantry modernisation.
- A high proportion of our armoured formations are holding vintage equipment. The fire support units, which form a part of our strike formations, do not possess matching mobility.
- Fire delivery systems need upgradation and a missile capability needs to be introduced in a greater proportion.

Recommended Force Structure: With the modernisation process, there will be a requirement for additional manpower for EW units, technical support units and back up logistic elements. However, with modernisation, the way ahead points to 'de-massification' of combat units, which retain the same effectiveness with upgraded state-of-the-art weapon systems. It is therefore imperative that the modernised combat units shed manpower on acquisition of the new weaponry and the manpower thus saved should be used for manning of force multipliers and other additional requirements in the restructured forces. A similar exercise would have to be carried out to cut out all non-essential logistic elements and a greater reliance on trade and existing national infrastructure introduced, for fielding additional technical support and back up elements.

The enhancement of the cutting edge capability of the Army, however, would greatly depend on the release of additional funds for modernisation, as also making up of existing voids. Non-availability of such funds would on the other hand gravely impinge on its combat potential and capability to deliver.

Problems of India-Pakistan-Kashmir relations are too complex to be resolved too easily. And it is Kashmir, which has strengthened the conflicting identities of two states, more so for Pakistan, which fears its total existence depends on the Kashmir issue only.

"India and Pakistan can work around this, but cannot remove it. ..... Kashmir is nowhere near as important domestically in either state as it was a number of years ago, and the military/strategic issues embedded in the Kashmir conflict could be finessed by introducing various verification and inspection regimes, agreement on force levels, pull backs etc...in addressing the issue of self determination of Kashmiris the two sides should be able to achieve an understanding over ground rules and context[4]."

But conclusion should not be drawn that challenges for the Indian army emanates only from her western neighbour. Border disputes with China, pressure of internal security duties, lack of sufficient financial support for modernization etc., will constantly trouble the army. And it would be the responsibility of the leadership to face them squarely.

Challenges for Indian Aerospace Power

Despite remarkable achievement of the air power in both Gulf and Kosovo wars, there have been speculations among the war strategists about the efficacy of air power and it's ubiquitousness. Critical analysis in recent years has brought out many lacunae in the theories of air power application. Yet, it can be said with certain surety that air power definitely paved the way for resounding success in Kuwait and in Yugoslavia. Does Indian Aerospace Power fit into such a scenario, is a question to be dealt most carefully. Indian Air Force neither has a clear superiority of force level in comparison to its two hostile neighbours, nor it can master the preponderance of force level in any foreseeable future. That focuses onto force structuring of IAF in the 21st century.

In addition, space segment of air power will be an integral part to its application in any future operations. Satellite surveillance, communication, and intelligence gathering will be the support expected from this new dimension. In case of India, this space segment had been in the hand of Indian Space

Research Organisation (ISRO), an independent Government of India undertaking. It has made impressive progress so far totally independent of any military application. Recently there attempts of certain convergence at a synergy between the Services and ISRO to bridge the gap to achieve better situational awareness. Over dependence on Global Positioning System (GPS) and such like foreign made facilities may be counter-productive over a period of time.

Aerospace power can be costly for a country like India. Underestimation may affect national security, whereas overestimation would be a drain on national resources. Force structuring would not be an easy task. !t would always be fraught with dangers of under/over estimation, and that could have severe impact on national security and economy. Therefore, to strike a balance between too much and too little, following factors are required to be analysed:

- Comparative evaluation of IAF with Pakistan Air Force (PAF) and People's Liberation Army and Air Force (PLAAF).
- An assessment of the present force level.
- Indicate inadequacies of the present combat fleet.
- And finally derive at an acceptable force structure.

Comparative Evaluation

Both China and Pakistan are concentrating on qualitative modernisation in their combat force levels. It may be stated that by the first quarter of the 21st century, the potential of PAF and PLAAF would have been enhanced considerably. Having identified the shortcoming of the PLAAF, the Chinese have endeavoured to achieve a more balanced force level. Su27s of which it has already procured about 97 and also procured 20 out 40 Su30sMK[5]. It is estimated that China is in the process of building up its air force most assiduously and will become a very potent threat 10 years hence. Chinese official defence budget was set at $15 billion in 2000, which was supposed to offset the loss of income from commercial ventures, and pay and allowances. But

SIPRI estimated that level of Chinese military expenditure has always been about 70-80 percent higher than official budget[6]. This explains the Chinese modernisation thrust. It is in the process of acquiring air to air refueling capabilities based on IL 76 aircraft and also is in the process of integrating AWACs in its force structure. Such rapid growth supported by a robust economy will pose a serious challenge to IAF in this century definitely.

It may be pointed out by the antagonist of airpower that Pakistan defence expenditure only stands at $3.5 billion in 2000 as against India's $14.4 billion. But the ground reality is that the IAF is rapidly losing the edge over PAF due to rapid obsolescence of its majority of force based on Mig 21s / Mig 23s and Mig 27s. These forces are no match to PAF's F16s. The USA stopped the delivery of the fresh purchases of these advanced jet fighters to Pakistan. But due to rapid changes in geopolitical situation in the post-Afghan war and permanent presence of the American troops on Pakistan soil will bring about changes in military dimension not entirely favourable to Indian Air Force in future. IAF will have to cater to these changes. Induction of one SU-30 squadron with few more in the pipeline will not be sufficient to maintain the edge over PAF for an indefinite period.

Objectives of IAF

IAF will never be able to muster overwhelming superiority of forces vis a vis its northern and western neighbours as was achieved by the coalition forces in Gulf and Kosovo wars. Yet it will be called to fight a war and apply its air might to force a decision on the enemy. There is definite requirement of the IAF to carry out balanced force structuring depended on the following objectives:

- Win against Pakistan and dissuade China.
- Defend against China and hold Pakistan.

Taking into account these factors, a generalised force structure for the Air Force can be conceived. The main emphasis should be on multi-role aircraft, modern sensors/ systems, force multipliers and a very good transport air fleet to support the air operations. The objective would be to carry out a successful joint air land battle.

Role of IAF

The role of IAF needs to be determined keeping the above objectives in view. The characteristics of the air power are described as, flexibility to adapt to changed circumstances, mobility to apply military power with speed and at long distances, concentration of power in time and space, and the ability to carry out offensive action deep inside enemy territory. This naturally translates the role of the Air Force as follows:

- Defence of the nation against attacks through the medium of air and space.
- During war, control of air and space to the required degree, to permit freedom of operations to own forces.
- Operation in mutual support with Army and Navy.
- Strategic Nuclear Deterrence: IAF was well suited for providing immediate retaliatory response in the event of a nuclear attack, till such time Indian Army took over the control of the Agni missile forces. These will no doubt spearhead the nuclear deterrence. And it is not understood what will be the role of IAF and flying assets. IAF capability of strategic nuclear deterrence has to take a back seat. It is yet to be seen how the strategic nuclear forces with its command and control system are installed in the changing Indian strategic environment.

Factors Affecting Force Structure

It may be repetitive to carry out the threat assessment again at this juncture. An elaborate threat assessment has been carried out in the previous chapters of this book. Therefore, the force structuring assessment would be carried out with such background and with such potent background to meet the tasks of the Air Force as enumerated above, while keeping in mind that force structure needs to comprise several complementing but disparate components.

Identification of the leading factors is not difficult, but achieving the right balance between them is extremely complex. There is no set pattern because the level of perceived threat or threats facing a nation cannot be numerically scaled. To work out a force structure, we have to examine the issues concerning diverse requirements of an air force, keeping in mind that its flexibility is its greatest strength. The arguments on quality vs. quantity will be unending. Nevertheless, the basic guideline in equipping an air force is the need to retain the flexibility and versatility of air power.

Hi-Tech vs. Numbers: For a modern air force, cutting edge technology is critically important. Yet, the impact of numbers cannot be overlooked. A large unsophisticated structured force could overwhelm a small technically superior force. This implies that, given the right circumstances, quality could be defeated by a large quantity. Also, it should be taken into consideration that whatever may be the sophistication, there is bound to be attrition even against an enemy devoid of modern warfare equipment. The danger of having a limited force size is that even moderate attrition rates could rapidly make an air force non-operational, even if it had been able to achieve its aims in the beginning of a conflict. Moreover, there will always remain a degree of uncertainty about the enemy threat and the anticipated attrition rates. In the case of the Indian Air Force, under emphasis of numbers, in view of the large forces that can be fielded by China, can lead to a problematic scenario. Therefore, while planning a force structure, such contingencies where the IAF is faced off against overwhelming numbers must also be taken into consideration.

Financial Impact: Keeping the threat perception and military objectives in mind, a basic force level for the three Services could be worked out. This would be the bottomline force requirement. Modernization will remain the central theme of force structure development, and the effectiveness of the force will depend on the availability of funds for modernisation. If adequate funds are not available, it may be necessary to reframe some of our military objectives.

Defence Research and Development Organisation (DRDO): The DRDO is the only defence research organisation in the country. Therefore, its plans, programmes and productivity must be considered with special emphasis on the need for increasing self-reliance towards the modernization process, so that the full potential of the organisation is exploited, and its development is not only reactive towards service needs, but also proactive. Our progression towards self-sufficiency should not totally preclude urgent interim purchase ex import. Such urgent requirements must be identified carefully, so that provisioning action is taken well in time and optimum value for money obtained. At the same time, where indigenous capability exists, the Services need to insist that 'good enough' made at home is better than the best from abroad.

Strategic Nuclear and Missile Capability: The decision-making apparatus at Service HQs are presently not considering operational utilization of any nascent or likely nuclear capability. Quite obviously, if defence planners are to project force structures over the middle and long term, both nuclear and missile programmes, capabilities and limitations, as also the process of development, need to be factored in. Otherwise, the force structure projections could be incorrect or even overtaken by progress and developments in these fields.

Force Level of IAF

The present IAF force level comprises 39.5 combat Sqns. Of these 20.5 are strike Sqns, 18 AD and one EW Sqn. In terms of technology, of the 20.5 strike Sqns, 9 are low tech, 11 are medium tech and only half a Sqn is a hi-tech. Similarly, out of the AD Squadrons, 14 are low tech and four are hi-tech. Thus

97% of the strike force is a mix of low and medium tech. Overall, only 11% of the IAF combat fleet is in the hi-tech category.

Inadequacies of Present Combat Fleet

The existing force level of the IAF is considered incapable of meeting the anticipated roles and missions through the year 2025 because of the following deficiencies:

(a) The combat fleet is an assortment of numerous types and version, with the consequently stupendous problems of operationalisation, training, maintenance and logistics. A majority of the fleet would only temporarily help in sustaining the existing force levels.

(b) Inadequate low level radar covers for effective air defence.

(c) Lack of adequate long and short range multiple role combat aircraft (MRCA).

(d) Non-availability of force multipliers like FRA, AWACS, ARM, UAV, etc.

(e) Limited all weather strike capability.

(f) Inadequate ECM suites.

(g) Lack of ECM resistant AD weapons, and secure communications.

(h) Inadequate standoff weapons, and secure communications.

(i) Obsolescence of ELINT.

(j) No helicopter EW.

(k) Inadequate C4 I2 capability.

(l) Inadequate WWR holdings.

(m) Inadequate transport/helicopter airlift capability to meet the Army requirements.

Assessment of IAF Capability

Combat Squadrons: It has already been predicted that by the year 2005-2006, IAF will reduce considerably from its present strength of 39.5 combat squadrons. That will erode IAF superiority, to such an extent that it will just match the PAF strength in another five to six years time whilst conceding overwhelming numerical superiority to China. If this trend is allowed to continue, air power projection capability of the IAF will be predicated on serious and undesirable voids. Therefore, IAF must strengthen itself by induction of the SU-30 class and replace combat obsolescence by upgrading aircraft and aircraft systems. It must provide the requisite fillip to development of the much-touted LCA, in order to retain qualitative and quantitative superiority vis-à-vis Pakistan and convey the unambiguous impression that China will not be in a position to dictate terms without severe penalty.

Heavy and Medium Transport Aircraft Squadrons: IAF must have the capability to airlift at least one mobile Brigade in one wave by the heavy lift transport aircraft. Considering the vast expanse of the nation and contingencies arising during peacetime and to maintain forward locations during war, it will be mandatory for the IAF to enhance the Medium Transport squadrons. The limited capability and strength of this fleet have been stretched to their limits. Unless hard decisions are taken to correct this situation, the much-needed flexibility of air power will be severely affected.

Medium Lift Helicopters: The IAF strength of medium lift helicopters is on the decline. Considering their role and their utility in the inhospitable terrain of the North and the Eastern jungles, it is necessary that steps should be taken for replacement of the older fleet. Moreover, this fleet has rendered yeomen service to the nation during all natural/national calamities and should be able to continue to do so. Overall strength of these forces is not sufficient to meet the peacetime requirement, let alone the wartime needs. Addition to these forces, apart from the replacements, would be an operational imperative for the Air Force and the nation as a whole.

Attack Helicopter Squadrons: Attack helicopters have become an integral part of any land operation. It would be prudent for the IAF to procure sufficient number of attack helicopters to support the three strike corps effectively.

Forces Multipliers

- **AWACS**: AWACS has become virtually synonymous with force multiplication. The strike force guided by AWACS and providing early warning can enhance the armament load carrying capability of each strike. Otherwise, these forces always pay a heavy penalty in terms of war load, to accommodate self-protection load. The AD capability of the Air Force will also achieve a quantum leap by receiving much needed airborne detection, control and guidance. Therefore, the IAF must procure AWACS to tackle any future conflict.
- **In Flight Refueling Capability:** In flight refueling capability will allow the IAF to disperse the valuable combat squadrons to bases generally not vulnerable to enemy air attack, in addition to enhancing their radius of action. Presently, intensity and congestion of air operations in the forward Indian bases are beyond desirable limits.
- **UAV/UCAV**: Unmanned Aerial Vehicle or Unmanned Combat Aerial Vehicle has not only the attraction of keeping combat casualties low but overcomes many human limitations. UAVs will be smaller, faster and more agile than manned fighters. Such unmanned vehicles could be used in preference to manned flights in a dense air defence battle space. These will be very useful to keep long vigil on the enemy activities and pass information to the command post. These can meet the operational requirements and fit into a sound plan of overall air operations. This was proved adequately during the Kosovo

war and more recently concluded Afghan war when Predators apart from doing surveillance were armed with Hellfire anti-tank missiles.

- **SSMs**: The Agni I and Prithvi class surface-to-surface missiles will add to the strategic strike capability. Induction of these missiles with a declared range of 700/250 kms will help to soften hard targets and relieve pressure on manned flights to ingress the intense AD environment of the adversary.
- **Precision Guided Munitions (PGMs)**: The devastating capability of PGMs has been adequately displayed during the Gulf war and recently concluded Afghan war. It would be essential for the IAF to use the Precision Guided Munitions from airborne platforms, to enhance strike success by several degrees of magnitude and exponentially reduce own attrition through standoff capability.
- **Joint Direct Attack Munitions (JDAMs):** In recently concluded Afghan war it was revealed that US AF has used the JDAMs effectively. These are less accurate than the laser-guided bombs but much cheaper. A $ 20,000 kit transforms an unguided iron bomb into a smart weapon, which uses the satellite linked positioning system to guide itself to the target, and can be dropped from varieties of platforms.

Sensors: Advances in sensor technology are increasingly creating opportunities to coordinate the operations of individual sensors in ways that produce significant operational benefits. IAF Air Defence architecture may not feel particularly confident about its prevailing sensors, which could face hostile challenges in this century unless particular attention is being paid to update the same. However, the sensors also are facing certain challenges as mentioned here:

- There is a requirement of precise registration of individual target signatures within a given time frame, both for accurate location and track determination.

- Sensors should support correlation and fusion of outputs from multiple sensors.
- Discrimination and sensitivity in a multi-target system, and detection of small targets in highly cluttered background.
- Retain accuracy and allow authentication of target data and continuously measure electronic environment parameters to assess the quality of target data.

Development of this century sensors are addressing the above and will have profound effect on air operations. Some of these are discussed below:

- **Ultra Wide Band (UWB) Radars[7]:** UWB technology uses highly sophisticated processing techniques. The principle is to use a wider spectrum of the available bandwidth. The transmission can be very low powered, because intelligence is transposed over a large band of frequencies. "UWB radars can be miniaturised due to very low power requirement and a large band width that permits use of simple RF generators"[8]. This would have serious effect on future data collection, to boost air defence capability.
- **ESM System[9]:** "Tamara built by Tessla Corporation may be one of these systems, which could detect the Stealth Aircraft like F 117. This system detects aircraft position by monitoring RF transmissions from the aircraft and then triangulating the returns".
- **IR Detectors[10]:** Air Defence Alert Device (ADAD) developed by Pilkington Systems of UK has the ability to lock on to an F117 beyond visual ranges.
- **Passive Cohesive Locator (PCL)[11]:** " It detects the perturbations in the background noise by the passage of an object. The atmosphere is saturated with RF energy from commercial broadcasters. A PCL system makes use of the reflection of this RF energy to detect a target. ...Lockheed Martin of the USA has successfully tested their PCL. The project called Silent Sentry matured after 15 years of research.... Its performance in recently

concluded trials was found to be the same like any Early Warning radar."

Electronic Warfare: IAF capabilities in the field of EW have to be assessed particularly considering the proliferation of equipment and tactics in this filed. Major effort would be required to study, understand and applicability of the following:

- Radiation detection and sorting in a dense signal.
- Standoff jammer for wide area penetration of enemy's electronic system.
- Extension of electromagnetic spectrum to milimetric and optical wavelengths.
- Warning of illumination by IR, Laser and Radar.
- Electronic protection of command and control systems.

Electromagnetic Pulse Bombs (EMP): The electromagnetic bomb generates electromagnetic shock wave and the energy produces such a storm of electromagnetic field that it produces short lived transient voltage of thousands of volts (Kilovolts) on exposed electrical conductors like wire or conductive tracks on printed circuits boards where exposed. This will cause serious disruption of information flow of the enemy. The idea would be to immobilise the enemy before destroying him. Such a weapon can be delivered by cruise missile or by any standoff platform and can be exploded within yards of a target. The explosion would generate high power microwaves, which could disable electronic circuitry in computers and communication equipment.

Deployment of EMP bombs is round the corner. Many reports emanating from various quarters suggest that EMP bombs have been perfected and will be deployed in future conflicts of this century. IAF will have to take this silent revolution in warfare into account. Procurement /development of this new weapons system has to be worked out and to be included in IAF doctrine.

Directed Energy Weapons (DEWs): It is better understood by the IAF top brass that a complete paradigm shift in military aspects would be expected when the Laser and Particle Beam weapons would be added to the military inventory. The cyberspace-based technologies supported by the DEWs in a future conflict might inflict unacceptable damage sufficient to alter the course of the battle.

Strategic Depth: IAF lacks strategic depth. Forward airfields are highly congested and are well within the striking range of PAF and even Pak artillery. This is extremely disadvantageous in terms of physical security of the aircraft and operational security of all missions. The enemy could monitor every movement on an airfield, which will be detrimental to mission completion. Strategic depth needs to be created urgently to provide security to IAF assets. Addition of few airfields at depth will add to better housing of costly combat aircraft and provide enhanced security for them.

Space Segment of Air Power

While talking about the importance of space and its impact on RMA, it was stated that," Space will be the crucial dimension for all these new weapons, for intelligence, communications, command, and targeting. And there is no doubt that space is surely destined to become a battlefield as land, sea and air have been through history. That is why it is expected that 21st century will witness another kind of warfare not known before[12] "

In the context of aerospace application, better situation awareness would be the most important factor to have edge over an enemy. That may be the reason for the Indian armed forces to come to an understanding with ISRO and plan the future defence application satellites as indicated below:

- **Technology Evaluation System (TES):** Launched in October 2001, with a single PAN camera and will have Resolution 2M. Coverage 14X 225km. Revisit five days. Could be improved to two days by lowering the orbit to 501km. Initial altitutde 562km.

- **Cartosat 1:** Planned launch early 2002. Resolution 1m. Sensor 2 Pan cameras with Swath 60km X 930km each camera to give stereo coverage.
- **Cartosat II:** Planned launch early 2003. Resolution 1m- Sensor 1PAN camera. Swath 15-20km (snap shot mode).
- **Resource SAT:** Scheduled launch 2005. Resolution 3m. Sensor Single multi-spectral camera. Swath 25km.
- **Space Based Surveillance (SBS):** Scheduled launch 2005. Orbit 560km X 281km. Resolution 0.5m. Life 1 ½ years. Swath to be finalised.

Coordinating agency for space application for the three Services would be Defence Image Processing and Application Centre (DIPAC), operating under Chief of Defence Staff (CDS) as and when it is instituted.

These may not be enough to become a viable aerospace power in the 21^{sT} century. It would be necessary for Indian military to exploit all space facilities and reduce over dependence on systems like Global Positioning System (GPS). Presently there are not much alternatives to American system orbiting around the earth. But adopting the system by the three services without thinking about an exit route could cripple any future operation, in the eventuality of its failure or even a partial disruption of the services. This would have severe repercussions on an on going battle and create situations unacceptable for many real time operations. In addition, following factors need to be addressed at the earliest:

- **Access to Space**: Strategic satellite reserve needs to be created for utilisation at the time of emergency. Military must have an easier and quicker access to space depending upon the operational requirement, rather than waiting for the ISRO to set the time schedule for each launch.
- **Observation and Situational Awareness:** For situational awareness, it would be necessary to convert the sensor data to information and knowledge, and move the data

when and where needed. Such activity may be conducted through land or space. But it is quite clear that space is going to be the primary medium and would dominate the 21st century battle space.

- **Space Control Technologies:** It is not clear whether India would be in a position to acquire sufficient technological base to control space to an extent that it could defend own space based resources and to disrupt, degrade, deny or destroy that of enemy. But it should be clearly understood that the countries like the USA and Russia have such capabilities. And China's growing space power is a clear indication towards achievement of this goal.

The entire discussion focuses on the factor that aerospace power now has the ability to carry out much more than what was understood before. It could carry out the bulk of the responsibility for beating down the enemy's fielded forces provided there is required strength, and application theory is correct. Thus it would help the other force elements to achieve their goals with a minimum of pain, effort and cost. While talking about the same, Jasjit Singh pointed out that:

"Under the impact of new technologies the capabilities, scope and role of air power are rapidly expanding; and with this expansion of its role and importance air power would play the primary, dominant and leading role in modern warfare. Any concept of air power playing a supporting role is not only redundant, but could be disastrous in war. Air power of today (and even more of tomorrow) is not only the major partner of military power of a nation, but is being pushed by technology into the role of greater leadership amongst the three main elements of military power[13]."

It also can be argued that air power with its vital adjuncts like surveillance, reconnaissance, combat platforms, PGMs and the mobile assets needed to deploy them, has changed the way major wars in the future will be fought. It was again pointed out that,

"Understanding and appreciating the impact of new technologies may also demand critical look into conventional wisdom and single-service prejudices, preferred positions, and hallowed perceptions regarding doctrines[14]."

Aerospace power of India will have to live up to the reputation generated by the recent wars in Gulf and Kosovo. Nothing less will be accepted by the strategic analysts despite the severe flaw that exists between the force level and the assessed threat of India. That should be the reason enough for the Indian Air Force to build up the force level supported by a sound financial plan. There is no doubt that a balanced force structure of IAF coupled with information power will have the potential to change the course of any future battle of the subcontinent.

Conclusion

Despite living in an intense nuclear environment it is clear that India will have to retain its superiority in conventional forces. Pakistan has finally seen the futility of an arms race with India and raises the nuclear bogey whenever she is cornered politically. There is concern with the strategic community that Pakistan being what it is now, could carry out nuclear attack on India despised by international community. And it considers its nuclear arsenal as its thread of survival. Pakistan is in great turmoil with President Musharraf at the helm of affairs. The visible rot is too deep. The Jehadis are not under the complete control of Pak army who fostered them for many years to bleed India. Al-Qaeda and its network has shifted from Afghanistan to Pakistan. And it is expected that the bursting of Pakistan will impose another unnecessary war on India. India is a stable democratic country following certain international norms which cannot be entirely expected of Pakistan. Therefore India has to be ready to go to war with Pakistan in future.

Battlefield scenarios have changed considerably as demonstrated in Gulf war, war in Kosovo and the most recently concluded Afghan war. Air land battle will be fought at a level not known before. Parallel wars will supersede serial warfare, number of sorties per kill will be replaced by "kill per sortie." Sensor to shooter loop will be more condensed. Advent of new weapon systems, sensors, and ascendancy of air and space power will have to be taken into operational plans. And joint doctrine will be necessary to fight the network centric air land battle of this century.

Notes:

[1] Lord Wavell.

[2] Ashley Tellis, "Stability in South Asia," Rand Report (Prepared for the United States Army).

[3] George K. Tenham, "Indian Strategic Thought," RAND R –4207-USDP (Prepared for the Under Secretary of Defense for Policy) p60.

[4] Ibid, p100.

[5] The Military Balance 2001-2002, p190.

[6] SIPRI Year Book 2001-2002 p255.

[7] Wg Cdr SB Deo, "Air Defence in The Coming Years," *Indian Defence Review*, Vol 16 (3) Jul-Sep 2001, p37-38.

[8] Ibid p38.

[9] Ibid p38.

[10] Ibid

[11] Ibid

[12] Bill Gertz, *The Washington Times*, February 23, 1999.

[13] Air Commodore Jasjit Singh, *Air Power in Modern Warfare*. 1985. Lancer International presentation, New Delhi, p270.

[14] Ibid

12

Challenges For Indian Navy

Jutting in the Indian Ocean India is placed in a unique position of maritime and geo-strategic importance. It oversees the vital choke points in the Indian Ocean that ensure passage of trade and energy resources to the energy demand heartland in Eastern and North Eastern Asia. In such geo-strategic circumstances the role of the maritime forces assumes enhanced importance. The Indian Navy is the fourth largest Navy in Asia and one of the primary navies operating in the Indian Ocean. It has the task of carrying out numerous functions both during war/warlike conditions and more importantly during the longer peace intervals. Thus facing numerous challenges in the process of carrying out its objectives.

Naval Objectives

Navies the world over have three primary functions. These include military mainly warfighting, followed by constabulary or policing and finally diplomacy in its coercive or non-coercive format along with power projection In nuclear weapon states, nuclear deterrence is another main objective of the navy. While warfighting is confined to war/ warlike situations, the next two functions are relevant in peace or less than peace conditions.[1]

Indian Navy not only has to defend a long coastline of 7,516.6 kms but in addition protect the 358 islands of Lakshawdeep and 723 islands of the Andaman & Nicobar chain from hostile maritime activity. Admittedly the present constabulary duties of the coast line against poaching or unauthorized fishing, drug/arms running and illegal immigration from neighbouring states has been entrusted to the Indian Coast Guard.[2]

Important diplomatic functions of power projection and showing the flag in peacetime are the other objectives of the Navy. They form an extension of the declared foreign policy objectives of any country. The recently held International Fleet Review 2001 at Mumbai was well attended by navies of the world and showcased the maritime might of the country. It also sent a subtle message of the coming of age of the Indian maritime power in the region and hence was an ideal example of maritime diplomacy and power projection during peace times.[3]

Apart from these objectives the most obvious and natural aim are the military functions of warfighting for sea control or sea denial.

The Challenges

In discharge of its objectives the Navy faces numerous challenges. These may be categorised under:

Geo-political: The geo-political challenges that face the nation from maritime dimension are many and varied. With Chinese PLA Navy making increasing forays into the Indian Ocean it is obvious that this can prove to be a strategic competitor to India. There is a considered view that this challenge is especially important in the background of the debatable hypothesis of encirclement of Chinese influence around India through its friends and allies. While this may be partially true, the fact remains that the Chinese are making dramatic and considerable efforts to modernise and streamline their navy to blue water status and thus be recognized as a maritime power. It flows as a consequence that soon the Chinese maritime might is likely to spill over to the Indian Ocean to secure their own maritime interests thus posing a challenge to the Indian Navy. This is likely to prove the most serious and significant challenge that the Indian Navy is likely to face in this century.

In comparison the challenge posed by the maritime forces of adversarial Pakistan are unlikely to be of long term significance. The reason for this is not only due to the comparable size of the Pakistan Navy (that is in the process of being modernized to add considerable punch) but more because of the dominant continental mindset that is predominant with the politically ruling Pakistan Army that is unlikely to give its Navy its due importance.

Force Levels: The Indian Navy also faces the biggest traditional challenge of an asymmetry between the maintained force levels in relation to the threats perceived in safeguarding of overall maritime interests and national aspirations.

The progressively greater focus on the seas with the Law of the Seas, (UNCLOS III) having far reaching effects on the manner in which maritime wealth will be exploited by nations in the future has primarily necessitated the need for greater surveillance and policing of our waters. This factor is definitely an added determinant in quantification of the available force levels.

Historically speaking the Government of India (GOI) for the first time formally stipulated the force level of the Indian Navy in 1964. According to the Defence Plan 1964-69 approved by the Emergency Committee of the Cabinet (ECC)on 17 March, 1964, the Navy was to progressively build its force level to 138 ships and submarines, the major surface ship elements of which were to be two aircraft carriers and 30 destroyers/frigates.

Whilst in the 70s and 80s there was considerable induction of new ships, submarines and aircraft into the Navy, there was a very marked slow down in the growth of the Navy thereafter. Consequently, the rate of induction of new ships was unable to keep pace with the deinduction of ships due to aging. This has led to a net reduction in the force level and also resulted in an imbalanced force mix, with a greater reduction in Fleet ships i.e. destroyers and frigates vis-à-vis other ships.

In consonance with the roles and missions discussed earlier, the Indian Navy needs to acquire a balanced force comprising the following:

Aircraft Carriers: Aircraft carriers are floating airfields that contribute significantly to achieving sea control, power projection and facilitate command and control functions. Despite their high capital costs, that fuels an intense military debate regarding their viability the fact remains that an aircraft carrier is the most potent military symbol for a maritime nation. The utility and usefulness of the carrier has been amply demonstrated in crises, hostilities and even in coercive diplomacy. The best example of this being its use during the Indo-Pak 71 war in the Eastern theatre.

The Indian Navy at present has only one aircraft carrier in its fleet- the aging Viraat. Given the present disposition of Eastern and Western fleets the requirement of additional one or two carriers is essential. While the proposed building of the indigenous ADS ship at Kochi shipyard may take atleast a decade or less, as an interim arrangement, the Kiev class Admiral Gorkshov offered by the Russians[4] is probably a welcome measure though the deal has been hanging fire due to prolonged price negotiations.[5]

Destroyers/Frigates: These ships form the very backbone of any Navy and are invariably its largest component. The Indian Navy presently has only eight destroyers, nine frigates and 24 corvettes.[6] Since a ship takes 8-10 years to build and has a life of about 20 years, the challenge is not only to foresee future force levels after analyzing determinants like future threat perceptions but to place orders well in advance. The series construction order must be given to shipyards in such a fashion so as to offset the phasing out (decommissioning) of older ships. Unfortunately this induction/deinduction process has not been linear and this has created vast gaps in force level build-ups leading to gaps during the past two decades.

Submarines: The importance of conventional submarines especially with context to our region cannot be under stressed. Indian Navy has operated the age old Foxtrot class of Russian origin since decades and is well versed in all aspects of conventional submarine operations. Presently the Navy operates 17 submarines that are mainly of Russian origin though four of the boats are of West German origin.[7] Seeing the current threat perceptions it is essential that Indian submarines likely to be purchased be equipped with cruise missile capability as is the case with the latest inductions.

Amphibious Capability: The responsibility of island territory protection coupled with the need for enhancing "operational reach and sea lift" necessitates the important requirement of having a well balanced amphibious force. At present, it is estimated that the Navy has the sea lift capability of approximately a brigade strength[8] with seven landing ships and ten LCUs. However with the phasing out of some older Landing Ship Tanks and Utilities the challenge for maintaining this capability remains in the face of its required enhancement given the present geo-political scenario and enhanced hostile maritime activity around the Island territories.

Surveillance: Optimum surveillance is a constant challenge that is faced by the Navy since it is of vital importance for carrying out any peacetime or wartime activity. There is a need to augment the number of long and medium range maritime patrol aircraft in the service and also induct UAVs for enhanced surveillance capabilities. This is because of the vast coastlines boundaries and subsequent EEZs require adequate surveillance monitoring platforms that are imperative for monitoring and intelligence gathering given the increased cases of piracy, gunrunning, flow of narcotics and other such activities in the seas surrounding the islands and the mainland. In this context, the requirement of the induction of Airborne Early Warning System for maritime purposes is also a necessity.

Security of Sea Lines of Communication

With nearly 89% of India's oil coming by the sea route it is but inevitable that the sea lines of communication (SLOC) for such extensive movement be secure. This security angle is enhanced manifold since most of the oil originates from West Asia and the SLOC pass through the areas of likely influence of the adversarial state of Pakistan. Added to this is the political turbulence of areas in the West Asian region that often makes the supply of oil from the region its hostage. In the past, supplies from this region have been disrupted at least on seven different occasions, all of which were due to political causes, and none were market driven.[9] The challenge for ensuring smooth energy supplies is a vital challenge for the IN. The recent effort in forming joint SLOC patrols on a multilateral basis is the logical path of progression in the correct direction.

Maintenance of Maritime Good Order

Piracy: Piracy has become the bane of the modern seafarer. The numerous cases of reported and unreported piracy have led to considerable concern and multi-national efforts required to control this violent menace which seems on the upsurge.

According to the World Maritime Piracy Report, SE Asia is one of the two areas that has been most affected by piracy[10]. This has probably a lot to do with not only geography of the area but also economic condition and the mindset of the people staying on the coast and hundreds of minor islands which dot the Malacca Straits and the South China Sea. In the recent past piracy related incidents seem to have spilled over these two areas into the Bay of Bengal and the Arabian Sea and hence there is a strong likelihood that the centre of gravity of piracy shifting to the waters around India. The most recent case of the *Alonda Rainbow* which was captured by the Indian Navy after a chase of 300 miles off Mumbai on 16 November, 1999,[11] and the case of *MV Cordiality* that was captured by the Sri Lankan terrorists off the coast of Trincomalee are pointers towards this observation.

The statistics of reported cases of piracy issued by the International Maritime Bureau clearly indicate the rising levels of this phenomenon. As many as 107 cases of attack were reported in 1991, and by 1999 the figures had risen to 285. Correspondingly in 1991 there were only 55 cases of piracy reported in SE Asia that rose to an alarmingly 160 in 1999.[12]

Unless serious efforts are made by law enforcement agencies and navies of various states in cooperating to bring this piracy into control (as shown by the case of *MV Alonda Rainbow*) this phenomena will not decrease but spin out of control.

Narco-Terrorism: An important adjunct to drug trafficking is its deep nexus with terrorism. With profit margins running into hundreds of percent depending on the procurement and the dispersion level of the shipment, drug trafficking is by far the most lucrative way of generating funds to fuel the ever growing terrorist activities and fuelling insurgencies around the world. Thus narco-terrorism is indeed a major security concern for India and its major maritime dimension is likely to prove to be a major challenge for IN/ Coast Guard as it affects the basic security fibre of the country.

Gun Running : The symbiotic relationship between gun running and trafficking of drugs is well known and established as one feeds on the other. It is extremely difficult to control one without controlling the other. Gun running by sea is by far the safest bet of transferring arms and ammunition around the world while drug trafficking the most lucrative. As stated insurgent movements around the world, like the Karens and the LTTE depend to a large extent on drug money to fuel their movement and equip their army. This security challenge is especially ripe in the A&N Islands and in Southern India that poses a challenge to the Navy.

Security Against Mining : Mining is one of the cheapest forms of carrying out maritime warfare. Mines can be laid by both war vessels or by airborne means. Civilian aircraft or civilian vessels can also be used to carry out mining operations, which can, to a very limited extent be carried out even by non-state actors. There are 48 navies in the world that possess the capability to carry out detailed mining operations. In reality, it is not even necessary to carry out actual mining operations. The mere threat or a well-calculated disinformation campaign about the laying of a minefield can very well serve the purpose of deterring any merchant ship from entering any 'affected' channel or strait.[13] Indian Navy with 10 ocean going minesweepers must be well equipped to face this challenge. [14]

Sea Based Deterrence

The advent of weapons of mass destruction (WMD) has had a profound effect on the overall security environment in the entire region of Southern Asia. While the draft Indian Nuclear Doctrine has proposed the development of a classical nuclear Triad for India, at present the Indian nuclear delivery platforms are in a dyadic format of land based missiles and airborne platforms. However an analysis of the available strategic nuclear platforms of the NWS reveals that there is an increasing tendency for these countries to lay an overwhelming stress on the sea based leg despite its high capital costs and technological requirement. This is primarily because of the enhanced survivability factor accompanied by its capability to provide assured punitive retaliation.[15] Thus contextual to the background of emerging trends in nuclear triad deployment/development it is essential that India rethinks its options regarding the developing the classical triad making it advantageous to opt for an unequilateral triad architecture that lays a predominant stress on the sea based leg while retaining the other two legs. Thus in effect keeping the other two lower technological legs as it might not be economically, operationally or technically feasible to eliminate them while retaining the major dependence on the survivable sea based leg.

India should therefore actively pursue its indigenous nuclear powered submarine project and associated strategic support infrastructure, and at the same time place enhanced efforts in procuring nuclear powered from friendly NWS. In this context the Russian offer of leasing/selling the Akula (971) class SSN to

India is specially relevant as its acquisition will ensure a quantum jump in India's nuclear capability. It would also give the IN the much needed experience in operationally and technically handling such platforms - a process that was initiated by the induction of INS Chakra (Charlie I class)- a nuclear powered submarine taken on lease from the Russians in 1988 and subsequently returned in 1991.[16]

Conclusion

Given the unique geo-strategic positioning of the country and its vast coastlines, along with dispersed Island territories the Indian Navy has vast maritime responsibilities. Consequently the challenges of being one of the major navies operating in the Indian Ocean are also varied and numerous. On one hand they emerge from the present geo-political situation with the Chinese trying to emerge as a maritime power with their navy making forays into the Indian Ocean and thus raising the likelihood of becoming a strategic competitor. On the other hand, ensuring energy security for a nation like ours whose energy demands are steadily growing and any interruption in its supply could prove disastrous. The security challenges include enforcement of good maritime order at a time when piracy drug/gun running and narco-terrorism have stated to have serious consequences on the security environment of the nation. The requirement of advanced nuclear delivery platforms in form of the sea based legs of the triad in a nuclear neighborhood are some of the other challenges that are likely to be faced in this century.

Finally the Indian Navy, while keeping its focus on self-reliance, needs to progressively develop so as to acquire the following:

(a) Strategic Submarine capability with the development of the nuclear submarine, as the assured and viable retaliatory strike option of the nation.

(b) Two aircraft carriers/air defence ships to counter air threat in the IOR emanating from shore based LRMP/Strike aircraft armed with anti-ship missiles.

(c) At least 30 destroyers and frigates with multi-mission capabilities.

(d) Conventional submarine for interdiction and protection of SLOC.

(e) Minesweeping and mine-hunting capability to ensure timely clearance of areas of vital interest such as approaches to harbours/choke points.

(f) Long and medium range maritime reconnaissance aircraft with strike capability to forestall threats to ships and vital installations. As also other airborne EW systems.

(g) Amphibious capability for joint operations in defence of Islands territories.

(h) Special Forces for undertaking clandestine operations and protection of maritime assets.

(i) LND forces for the defence of own harbours.

(j) Fleet Tankers/Logistic ships to ensure sustained logistic support to ships at sea.

(k) Survey ships to enhance our capability in terms of oceanography and hydrography.

Further, there is a growing need to lay greater emphasis on future combat capabilities so that our forces enjoy technological advantage over potential adversaries. Central to this is the ability to gather, analyse and disseminate relevant information to the appropriate place at the opportune time. The crucial force multipliers that will be relevant in any future combat are:

(a) Intelligence, surveillance, target acquisition and reconnaissance (ISTAR).

(b) Command, Control, Communication and Computers (C4).

(c) Information Warfare (IW).

(d) Precision weapon systems with standoff capabilities.

(e) Stealth.

NOTES :

1 P K Ghosh, "Fleet Reviews and Goodwill Visits as an Extension of Foreign Policy", Commentary, *Strategic Analysis*, April 2001, Vol XXV, p.136.

2 The details of the statutory duties and responsibilities of the Indian Coast Guard are listed in Section 14 of the Coast Guard Act. Also see "The Coast Guard", Coast Guard Headquarters, (1993) p.1-2.

3 Ibid p.135.

4 For a detailed treatment see Rahul Roy Choudhury, *India's Maritime Security*, Knowledge World, New Delhi, 2001 p.134-135.

5 See Huma Siddiqui, "Pricing Holds up Adm Gorshkov Deal with Russians" *Financial Express*, New Delhi,March 15 2002.

6 Jane's Fighting Ships 2001-2002, p.295.

7 Ibid

8 Roy Choudhury, n.4, p.118.

9 P K Ghosh, "Maritime Security Environment in SE Asia and the Indian Ocean Region", (unpublished) Paper presented at the Tenth CSCAP Maritime Cooperation Working Group Meeting in May 2001 at Kuala Lumpur.

10 ICC International Maritime Bureau, "Piracy Report Reveals Major Increase in Attacks in 1999" available at Internet site< http//www.icc-ccs.org>

11 ICC International Maritime Bureau, "Piracy and Armed Robbery Against Ships" Annual Report 1 January to 31 December 1999, pp.12-13.

12 Ibid. pp.3,5,6,11.

13 P K Ghosh,n.9.

14 Jane's, n.6.

15 For a detailed treatment on emerging trends in nuclear triad see P K Ghosh, "Emerging Trends in the Nuclear Triad", *Strategic Analysis*, Vol. XXV May 2001.

16 Hans M Kristensen and Joshua M Handler, "Tables for Nuclear Forces" (Appendix 6A) SIPRI Year Book 2001, (Oxford University Press 2001), p.479.

13

Hijacking:

The New Age Menace

Terrorism continues to haunt the world. The trends in new international terrorism has taken a different turn since the 1990s which points a finger towards some groups of Islamic militants who have been mutating into a potentially extremely dangerous threat to international security than the groups that dominated the international scene in 1970s or 1980s. The direct fallout has been the bombing of World Trade Centre in 1993 by Ramzi Yousef and now the hijacking of four American airliners on 11 September, 2001. Hijacking four airliners almost simultaneously from different locations and crash into multiple targets within a matter of about 90 minutes surpasses every imagination of hijacking known so far. The attacks were amazing in their diabolicalness. The United Nations International Day of Peace was shattered watching the twin towers of the World Trade Center (WTC) collapse, killing thousands of innocent civilians. Osama bin-Laden expressed his happiness, but he also did not know that the heat from thousands of liters of aviation fuel would melt down the steel supports and collapse the trade centre. Mainland America never faced such an attack before. The peace was broken by the rudest shock of most heinous kind. No airline pilot could have been forced to fly into a sky scrapper even at gunpoint. But few educated and well-informed Islamic radicals learnt to fly this type of aircraft and flew into these obstacles. World Trade Centre collapsed; Pentagon became a victim to this batch of carefully groomed neo-terrorists. They perished along with hundreds of ill-fated innocent passengers of the aircraft. Thousands of Americans died under the rubbles of WTC and Pentagon. Completely brainwashed by Osama bin-Laden and his infamous Al-Qaeda organisation, these young men believed in their martyrdom and place in heaven.

While the number of international terrorist attacks has decreased, the lethality of the attacks has increased considerably. Suicide and car bomb attacks are preferred methods to achieve the goal. Palestinian attacks in Israel, the truck-bombs used in Nairobi and Dar-es-Salam against the American embassies in August 1998, October 12, 2000, naval suicide bombing against the *U.S.S Cole* in the harbor of Aden are the glaring examples of such kind of atrocities. On the other hand, the hijacking of airplanes has reduced which was the principal modus operandi of terrorist organizations in the 1970s and 1980s. These changes may be due to strict security measures taken by most countries at airports and prior to boarding an aircraft. But the September 11 incident shook up the entire world. Terrorism has been a tool in political or ethnic conflicts; and spread to unknown areas. The radical Islamic terrorism **threat** has grown, and we now see countries like India, China and Russia confronting this kind of terrorism. The number of states sponsoring or supporting terrorism has also diminished. Yet, Iran, Syria, Lebanon, and, to a lesser degree, Sudan, continue to present a threat to the international community. Lately Pakistan and Afghanistan became the biggest perpetrators of the radical Islamic terrorism, fostered by personalities like the ill famed Osama bin-Laden and his Al-Qaeda network.

The first real non-conventional terrorist event, the sarin gas attack by the Japanese cult Aum Shinrikyo in March 1995 in the Tokyo subway, had broken the taboo in the use of weapons of mass destruction (WMD), followed by 11th September attack on twin towers of WTC. No one ever could imagine that a passenger airliner filled with JP (jet propulsion) fuel could be used as WMD to kill a mass of population by committing hara-kiri. These have been very significant events in the history of modern terrorism. If we take into consideration the number of very serious incidents that have occurred till date, we get an indication that despite all the efforts, security and intelligence services, we failed to find answers to all such challenges posed by the new terrorists. Careful and meticulous planning, religious grooming, technological assistance and, powerful motivation to embrace martyrdom took the intelligence agencies by complete surprise on September 11, 2001.

Secret Plans

There have been numerous incidents of hijacking by a small but well prepared group of people, backed by right kind of logistics. But hijacking four aircraft on a well coordinated time schedule from different airports and crashing them into multiple targets within a matter of 90 minutes of each other is a different kind of operation. This act required logistic support, coordination involving hundreds of people and months of secret preparation. Strange kind of motivation resulted Mohmmed Atta and his compatriots to commit suicide with thousands of liters of jet propulsion fuel on board the flying airliners. This leads to the fact that the perpetrators of this act had to find five pilots, train them to fly the aircraft like Boeing 757/767 airplanes and indoctrinate them to die for a cause. The attackers selected airline planes on routine flights to avoid any early detection. Targets were selected carefully. Destruction of which would send the right kind of message to the most powerful country of the world. From the attackers point of view the entire plan was hatched in total secrecy, planned in careful details and well executed. It definitely required access to significant amount of information resources. How an attack like this could have been defended is the question raised after the attack. Suppose there were certain surface to air missile sites near New York or Washington and it was detected that airplanes had been hijacked and were heading towards the present targets, could the air defence commander have shot down the planes over a crowded city like New York?

- In earlier cases, it has always been the negotiation with the hijackers once the hijacked aircraft got on the ground. That idea is dead. Al-Qaeda has invented a new brand of terrorists who could die easily. Most difficult part of any hijacking had been getting away once the task was accomplished. But 11th September attack resolved that difficult part of hijacking. Al-Qaeda morons got away to heaven along with the airplanes and many others. May be in future they wouldn't mind flying into other sky scrappers. Generally the suicide bombers have been young, single, and fanatical and had nothing to lose. On the contrary, 11th September hijackers were older, well educated and more experienced. What FBI has revealed is that 50 of them planned and 19 were willing to die.

Hijackings on 11 September 2001

- **American Airlines Flight 11:** The flight departed from Logan-Boston for Los Angeles at 07:59 hrs. Five Arab terrorists hijacked the aircraft, who switched off the transponder and changed its course to New York. The area controllers saw the movement and tried to contact the aircraft also to warn other reciprocal traffic. It was seen to descend rapidly increasing its speed and finally crashing into the side of the 110-storey New York WTC North Tower at 08:45 hrs. The tower caught fire following an explosion and collapsed at 10:29 hrs. Presumably a member of the flight might have pressed a switch, which permitted the air traffic controllers to overhear at 08:28 hrs "don't do anything foolish. You are not going to get hurt. We have more planes. We have other planes." Details of the flight are:
  - o Type of aircraft: Boeing 757-223 ER
  - o Year built: 1987
  - o Total fuel carried: 43,402 Ltrs
  - o Operator: American Airlines.

- o Flight No: 11
- o Time of Take off: 07:59 hrs.
- o Time of Crash: 08:45 hrs
- o Crew on board: 11
- o Passengers on board: 81
- o Total Casualty: 92
- o Flight: Boston-Logan, MA to Los Angeles CA

- **United Airlines Flight 175**: Flight 175 departed again from Logan-Boston for Los Angeles at 08:14 hrs. Exactly after 15 minutes from American Airlines flight 11. Five Arab terrorists hijacked the aircraft. The hijackers switched off the transponder and headed for New York. Peter Hanson called his father from the plane and told him that the hijackers were stabbing the attendants in order to force the pilots to open the cockpit doors. At 09:03 hrs the aircraft was flown into the side of the New York WTC South tower, exactly after 18 minutes from the time the North Tower was hit. The tower caught fire and collapsed at 09:50 hrs. Details of the flight are:
  - o Type of Aircraft: Boeing 767-222
  - o Operator: United Airlines.
  - o Year of manufacture: 1983.
  - o Total Fuel capacity[1]: 91,379 Ltrs
  - o Time of Take off: 08:14 hrs.
  - o Time of Crash: 09:03 hrs.
  - o Crew: 9
  - o Passengers: 56
  - o Total Casualty: 65
  - o Flight No. 175: Boston-Logan to Los Angeles CA
- **American Airlines Flight 77**: The Washington-Dalles flight departed at 08:10 hrs for Los Angeles. The aircraft was hijacked in flight by the terrorists and they took over control and after switching off the transponder changed course to Washington. Wife of solicitor general, Barbara K Olson, called her husband, at the Justice Department at 09:25 hrs from the plane to tell him about the hijacking and to report that the passengers and pilots were held

back in the rearside of the plane. Last reported data (08:50 hrs) was that the aircraft flying at flight level 350 at 458 knots. The aircraft descended to circuit height approaching from north. The aircraft reported passing overhead the White House and entered a tight 270-degree turn, heading for the US Department of Defence (The Pentagon). It clipped trees and light posts before crashing into the south west face of the Pentagon. Followed by an explosion the building caught fire. Details of the flight are:

- o Type of Aircraft: Boeing 757-23
- o Operator: American Airlines.
- o Year of Manufacture: 1991
- o Total fuel capacity: 43,402 Ltrs
- o Time of Take off: 08:10 hrs.
- o Time of Crash: 09:40 hrs.
- o Crew: 6
- o Passengers: 58
- o Total casualty: 64
- o Flight No. AA 77: Washington-Dulles, NY-Los Angeles CA

- **United Airlines Flight 93**: The New York-Newark flight departed for San Francisco at 08:10 hrs. Four Arab hijackers hijacked the aircraft. In the passenger cabin three hijackers wearing red bandannas herd the passengers to the back of the plane. Passenger Tom Burnett called his wife to tell her about the hijacking, she in turn alerted the FBI and when he was connected again, he was connected to the FBI and described the hijackers. Jeremy Greek, another passenger, called his wife in New York and reported that three Iranian looking men had hijacked the plane, one of which had a red box around his waist, which they claimed to be a bomb. Jeremy asked his wife whether the WTC bombing was a fact as he heard from other passengers. Whereas Todd Beamer, another passenger, tried to place a credit card call but was routed to a customer service representative, who passed him on to his supervisor Lisa Jefferson. She called FBI. Beamer reported that one passenger was dead. Later he reported that the men were planning "to jump" the hijackers. The last words of Jeffersen were heard, " are you ready guys? Let's roll. " The cockpit sound recorder later found contained screams, presumably of hijackers, "get out of here," then screaming and other sounds and silence.

The aircraft changed course probably for Washington. At 09:35 hrs near Cleveland, the aircraft changed course to the south and later at 09:45 hrs to south west at fight level 350. At 10:10 hrs the aircraft crashed out of control southwest of Pittsburgh in Somerset county. There is speculation that the passengers' resistance as described before must have led to the crash before it could reach the intended target. There was every indication that the aircraft could have been flown into Air Force One or White House itself. Details of the flight are:

- o Flight 93: New York-Newark, NJ to San Francisco, CA
- o Type of Aircraft: Boeing 757-222
- o Operator: United Airlines.
- o Total Fuel capacity[2]: 43,402 Ltrs
- o Time of Take off: 08:10 hrs.
- o Time of Crash: 10:10 hrs.
- o Crew: 7
- o Passengers: 37
- o Total Casualty: 44

Know the Hijackers

Islamic extremists wish to turn back history and fight the Western modernity, which threatens their medieval theocracy[3]. They easily blend into the Western society and await their instructions to move. Bin-Laden's cohorts learnt the America's openness and technological superiority and discovered the effectiveness of turning flying airliners laden with thousands of jet propulsion fuel into weapon of mass destruction (WMD). Cell phones, encrypted e-mails only assisted them to keep in touch with their masters in

Afghanistan and receive instruction without raising any suspicion. They were not poor or desperate who were indoctrinated by a false promise of heaven through martyrdom. They were well educated, lived comfortably with sufficient money to rent cars, go to school and learn to fly at different US facilities. Two of the alleged hijackers of Flight 93 Ahmed Alhjaznawi and Ziad Jarrah drove a Ford Ranger[4] They used both the Naval and Air Force facilities to learn flying. The 33-year old Mohammed Atta, the mastermind behind the attacks, could issue a cheque of $ 10,000/- to learn flying in one of the Florida's flying training establishments. And as reported, he along with another person went to fly Boeing 757 simulator. They paid $ 1,500 each for six hours of flying training. But it was most surprising that they only wanted to learn to turn the aircraft. They were not interested to learn to take off or land, which is the primary requirement to fly an aircraft. This was unusual but none took it seriously, and no one thought of raising any report. Bin-Laden had been careful while planning his moves. Al-Qaeda had been dispatching the operators well in advance before an operation. These cohorts took time to blend themselves into the system and wait for the opportune moment to strike. In this case also the idea of using suicide pilots to hurt the USA must have been carefully planned a long time ago. The activating agents of Al-Queda must have arrived in the USA just about a week or two before the final act. It is not easy for an open society like America to defend against such terrorists who are educated, smart, patient and ready to die. Evil genius of Bin-Laden and his Al-Qaeda has created these morons successfully. And it is not known how many of them are roaming around freely and are ready to strike despite being flushed out of Afghanistan. Federal Aviation Agency (FAA), the controller of all civil aviation, was dismayed to plot the erratic behaviour of the hijacked airliners for about 15 minutes. The controllers were at a loss. They tried to raise the aircraft on R/T but failed to do so. They noticed the collision course of the hijacked aircraft and warned the others flying on that route. They noticed the aircraft taking sharp evasive action to avoid air-to-air collision. They did not suspect hijacking till such time the aircraft crashed into WTC. The US fighter planes did not arrive on nation's capital for another 15 minutes[5], Bin-Laden activists, Mohammed Atta and his 19 trained compatriots, had shattered the myth that "**America is safe and it cannot happen here**." Names of Mohammed Atta and his men:

- **American Airlines Flight 11:**
  - Waleed Alshehri
  - Wail M Alsheri
  - Mohammed Atta
  - Abdul Aziz Alomari
  - Satam Al Suqami
- **United Airlines Flight 175:**
  - Marawan Alshehhi
  - Fayez Banihammad
  - Mohald Alshehri
  - Hamza Al Ghamdi
  - Ahmed Al Ghamdi
- **American Airlines Flight 77:**
  - Kahlid al-Mindhar
  - Majed Moqued
  - Nawaf al-Hazmi
  - Salem al-Hazmi
  - Hani Hanjour
- **United Airlines Flight 93:**
  - Ahmed al-Haznawi
  - Ahmed Alnami
  - Ziad Jarrah
  - Saeed Alghamdi

Mohammed Atta: As identified by the US Justice Department, Mohammed Atta had been the mastermind behind the entire episode. On September 11, 2001, he was the one who led the first attack and crashed into the World Trade Centre. His skill to turn a Boeing, learnt in a Florida simulator came handy. He could avoid reciprocal traffic and guided the aircraft successfully to this target. His identity was initially confused with another Mahmoud Mahmoud Atta[6] involved in a bus bombing in1986 in the Israeli controlled West Bank. This Mahmoud Atta was a naturalized citizen of the USA but much older than the Mohammed Atta in question. The hijacker Mohammed Atta was born in Egypt in 1967 or 1968 who carried a Saudi passport also. After graduating in Architecture from Cairo University, he went to Germany as a student of urban planning and studied at the Technical University of Hamburg from 1993-1996[7]. His German friends describe him as an intelligent man with religious belief. He became more and more religious after a pilgrimage to Mecca in 1995. He began a religious prayer group in 1999 in the University and could have used it to recruit fundamentalists. In July 2000, he enrolled himself at Huffman Aviation International in Venice, Florida, by paying $ 10,000. Following this flying training he enrolled himself with the Boeing simulator to learn to turn a Boeing. Marwan Alshehhi, a hijacker of United Airlines 175, always accompanied him. Atta came back to Germany in May 2001, briefly and returned to Florida via Spain.

The US investigators found out that on September 4, 2001, Atta sent a package to Mustafa Ahmed in the UAE, the central contact point to all money transactions of Al-Qaeda. He was seen drinking and playing video games in a Florida sports bar week before the attack. He spent some time with another hijacker, Abdul Alomari, in South Portland and Scarborough, Maine.[8] On the morning of 11th September, they went to Portland airport, flew to Logan, Boston, and boarded the American Airlines.

One of Atta's baggages could not be loaded in Flight 11 and fell into the hands of the investigators. The bag contained airlines uniform, and a four-page document in Arabic. A similar document was found with the other terrorists of the hijacked aircraft. It contained a list of instructions like, "make an oath to die and renew your intentions. You should feel complete tranquility, because the time between you and your marriage in heaven is very short." This also instructed, " check your weapon before you leave and long before you leave. You must make your knife sharp and you must not discomfort your animal during the slaughter[9]."

- **Ziad Jarrah**[10]. The 27-year old Ziad Jarrah was born in 1974, and is suspected to be one of the hijackers of United Airlines flight 93. He like Atta also studied in Germany until the year 2000. He also left for Florida to attend a Flying training school. His girl friend in Bochum filed a missing person report. From Florida he wrote to his girl friend and due to wrong address the letter was returned to the USA and became an evidence for the investigators. He wrote, " I have done what I have to. You should be very proud, it is an honour, and you will see the result, and everyone will be happy. Hold on to what you have until we meet each other again." However his family in Lebanon claims that he was an innocent passenger on the plane or that possibly his is a case of mistaken identity. But he has not been seen after the attack.

- **Khalid al-Midhar.**[11] He was one of the five men identified by the US Justice Department as hijacker of the American Airlines flight 77. For the first time CIA got video of the suspect in a meeting of terrorists in Kuala Lumpur, Malaysia. He had connection with Osama bin- Laden when it was established that he attended meeting with a top Bin-Laden security official probably connected with *USS Cole* bombing. The US emigration was advised to look out for him and his close associate Nawaf al-Hazmi and prevent their entry into the US. But it was discovered later that both of them had entered the USA through Los Angeles National Airport in 2000 and then went to New York area in 2001. Al-Midhar rented rooms in San Diego from Abdus Sattar Sheikh.

Selection of Targets

The men who commandeered the airliners that struck the symbols of American power, the World Trade Center and the Pentagon, are the neo-terrorists created by Osama bin-Laden for whom the glory of martyrdom inspired suicide attack. Indoctrination of these young educated men led to establish their faith that one who dies a martyr for the faith goes directly to heaven and enjoys its rewards. They declared Jehad (the holy war). The word has many meanings. It includes the fight to control one's passions, to convert non-believers and improve Muslim society's moralistic views. Herein Gilles Kepel stated, " Once you say you are in a state of Jehad, then all the usual rules of society are suspended and the danger is that social structure will end in ruins[12] ."

Jehad is against this American capitalist power. To curb the ever increasing American influence Al-Qaeda under the guidance of Osama bin-Laden carefully planned the attacks on the US mainland considered invincible so far. The targets were selected carefully and the means of attack has been novel. But why Al-Qaeda selected these targets:

New York and World Trade Centre: New York is the world's largest financial capital, and World Trade Centre stood like the proud representative of the capitalist world. This is the largest city of United States and may be the most important city of the world. The United Nations Headquarters is in New York City, which may confer it the title of global capital. This city is the most densely populated in whole of the USA, with an eight million population in the City of New York and 20 million in the entire metropolitan area[13] . The city began with Dutch settlement in 1624 and was called Niew Amsterdam. It became the main city of the Dutch colony of Niew Netherland. In 1664, Britishers captured the city and renamed it as the city of "New York". For a brief period in 1789 and 1790 it was the capital of USA. Modern city of New York was created in 1898[14] .

- **World Trade Centre:** WTC consisted of seven buildings with the twin towers in the center of them. The centre was designed by American architect Minoru Yamasaki[15] . Each of the towers had 110 stories. The height of the towers was 417 meters (1368 feet North), and 415 mtrs (1,362 feet, South) respectively. The towers were completed in 1972 and 1973. At that time these were the tallest building on earth, remained as such till such time Chicago Sears' Tower was built with a height of 1450 feet. Each tower contained 240 vertical steel columns outside the building, called Vierendeel trusses, and these were bound to each other using ordinary steel trusses. There were dampers in the structure. This strong shell made it easy to build the floors with light steel and concrete with internal walls not needed for structural integrity. The tower was extremely light despite its size. A six level basement was built in the foundations. One of the worlds' largest gold depositories was stored underneath the World Trade Center, owned by a group of commercial banks. The estimated 1993 value was one billion dollars. Even the WTC bomb detonation near the vault could not break them open.

- **Pentagon[16]:** Designed by George Edwin Bergstrom the Pentagon shaped building houses the Department of Defence of the United States. It houses approximately 23,000 military and civilian personnel. It was built during the Second World War mostly with concrete and structural steel and has five floors. This is one of the world's largest office building with three times the floor space of the Empire State Building. It has 17.5 miles of corridors but it takes only seven minutes to walk between any two points in the building. American Airlines flight 77 crashed into the building on September 11, 2001, causing one part of the building to collapse.
- **Pittsburgh[17]:** Though Pittsburgh was not a chosen target by the terrorists yet a word about the same would complete this chapter. Pittsburgh has been an industrial town at the confluence of the Monongahela, the Allegheny and the Ohio rivers. It is the largest inland port of USA. Industrialists like Andrew Carnegie and Andrew Mellon built their fortune in this town. Pittsburgh was one of the largest steel producers in the world in 20th century. The city now houses industries for robotics, computer software, biotechnology and environmental engineering in place of

iron and steel. The population of great Pittsburgh is over 2.3 million. United Airlines flight 93 crashed southeast of Pittsburgh in Somerset county.

Terrorist Attacks (Prior to September 11)

In past there were also terrorist attacks, which rocked the contemporary world. But September 11 attack has surpassed every other attack in its diabolicalness and novelty. The casualty number has been overwhelming. Economic impact has been insurmountable. That is why it is said that Bin-Laden has scored point over America. List of incidents given below will prove the point discussed:

- 1972 Munich Olympic massacre
- 1972 Bloody Friday Bombing by IRA
- 1976 Orlando Letlier assassinated in Washington by Chilean Government
- 1985 TWA 847 hijacking
- 1988 Pan Am 103
- 1993 World Trade Center bombing
- 1995 Tokyo Sarin Attack
- 1997 Khobar Towers bombing
- 1998 US Embassy bombing
- 2000 *USS Cole* bombing

Hijacking Cases (Prior to September 11)

- Hijacking had always been the handiwork of a few disparate groups, with the intention of using its passengers as hostages to gain certain advantages from a particular government. It followed a pattern of negotiations and settlements if at all. 11th September incidents have changed the meaning of hijacking. The attackers of the WTC and Pentagon rewrote the hijackers' behavioural pattern. Al-Qaeda created this new brand of hijackers who were not so young, but well educated and ready to die to usher in the medieval theocracy. Previously the biggest problem of the hijackers was to get away without getting hurt once the mission was a accomplished. But current Al-Qaeda terrorists had solved the problem very neatly. They found the morons to fly into skyscrapers. The plan was perfect. The operation was a complete success. There was no leakage of information. These attacks have changed the world's conception of hijacking. During the last 30 years there were many cases of hijacking, but these were child's play in comparison. And that can be understood once we study the other hijacking cases given here:

- 1968: Three members of the Popular Front for the Liberation of Palestine (PFLP) hijacked an Israeli El-Al plane to Rome. The aircraft was then flown to Algiers. Negotiations extended for a period of 40 days. Both the hijackers and passengers went free.
- 1970: PFLP attempted hijack of four aircraft. They succeeded with three planes to fly to the Jordanian desert. The hijackers blew up the aircraft after releasing most of the hostages. Last of the group of hostages were released once the seven Palestinian prisoners were released.
- 1977: Palestinians hijacked a Lufthansa airliner to Mogadishu. German commandoes stormed the aircraft. Three hijackers were killed and 86 hostages were freed.
- 1981: A Pakistani International Airlines plane was hijacked to Kabul. One passenger was killed and the plane was flown to Damascus. After 13 days of negotiations hostages were freed in exchange of release of fifty political prisoners.
- 1982: Palestinians hijacked an Egypt Air aircraft to Malta. Egyptian commandoes stormed the aircraft and 59 people got killed in the gun battle.

- 1985: A TWA flight hijacked by Lebanese Shi'a gunmen to Beirut with 153 people on board. Israelis freed 31 Lebanese prisoners in exchange of the hostages.
- 1986: A Pan Am flight carrying 400 passengers landed at Karachi with hijackers on board. Pakistani security forces stormed the aircraft killing twenty-two people.
- 1990: A Chinese airliner was hijacked which crashed while trying to land in Canton killing 128 people.
- 1994: A Air France airplane was hijacked by four Islamic terrorists to Algiers. It was flown to Marseilles where the French commandos killed the hijackers and freed the 170 passengers.
- 1996: A Ethiopian airlines crashed into the Indian Ocean after the hijackers forced a crash landing. 50 passengers survived but 125 died.
- 1999: A Indian Airlines aircraft was hijacked by Kashmiri militants and was diverted to Kandahar. Indian Government agreed to release three top Pakistani militants in exchange for the hostages.

Hijacking and International Law

Hijacking of international flight of a civilian aircraft is covered by principles of both international customary law and treaty law. General norms are:

- That every state has an obligation to ensure that its territory is not being used by any person in any manner to the detriment of another state.
- That state owes an obligation to the international community to ensure protection of human rights.
- Contemporary international law underscores the duty of every state to:

- o Prevent perpetration of terrorist act against another state or innocent individuals.
- o Consequently to cooperate with other states in preventing and combating terrorists and in punishing terrorists.

There are four multilateral treaties directly relevant to aerial hijacking. The first three were evolved under the aegis of International Civil Aviation Organisation (ICAO). The UN General Assembly adopted the fourth. These are:

- Tokyo Convention on Offences and Certain other acts committed on Board the Aircraft, 1963[18].
- The Hague Convention for the Suppression of Unlawful Seizure of Aircraft, 1970[19].
- The Montreal Convention for the Suppression of Unlawful Act against the Safety of Civil Aviation, 1971[20].
- The International Convention against the Taking Hostages, 1979[21].

The Hague and Montreal conventions defined the offence of hijacking and stipulate that an attempt to commit that offence or to aid and abet its commission also amounts to hijacking. The conventions like any other anti-terrorism conventions recognize explicitly or by necessary implications the jurisdictional claims of at least seven categories of states[22]. These are:

- The state where the aircraft is registered.
- The state of operator of the aircraft.
- The state in whose territory the aircraft lands with offender on board.
- The state in whose territory the alleged offender is found.
- The state in whose national is the alleged offence.
- The state in whose territory or in whose airspace the offence was committed.
- The state whose security is directly affected by the offense.

As direct consequences of the possibility of competing jurisdictional claims, the Conventions for the first time established a general obligation on any party in whose territory the offender is found either to take measures to prosecute him or to extradite him. All states party to the Conventions have an obligation to render the prosecuting state the greatest measure of assistance in connection with criminal proceedings including evidently handing over of any evidence or other information in its possession.

The convention recognize the supervisory role of the Council of the ICAO by requiring each party state to make a full report to the Council at the earliest. The reporting obligations are mandatory. It may be noted that Chicago convention of 1944 has given special powers to ICAO[23]. Under its constitution, it is to strive for the safety of civil aviation the world over. The convention also provides for an optional settlement mechanism of arbitration or access to the International Court of Justice[24].

Prevention of Hijacking

Aviation industry has grown in leaps and bounds, despite impediments like occasional accidents and stray incidents of hijacking. But the tragic incident of 11th September has made the international aviation community to sit down and think how safe are airlines from the hijackers. In India there have been several measures in vogue prior to departure of any flight, both national and international. But no one knows how to deal with the Al-Qaeda type of terrorists, who are taught to commit hara-kiri. To date none of the measures provide adequate protection from an attack like the one on 11th September. It may be an interesting point to note that there had been no incident of hijacking of Israeli aircraft so far, despite their continuous operation in and out of a most sensitive area of the world, apart from the only incident of El Al hijacking mentioned before. The Israelis implemented certain measures like reinforced and locked doors between cockpit and the passenger cabin, place all cargo in decompression chambers before take off, to trigger bombs set to sense altitudes. These may be expensive but has proved their worth.

Sky Marshals[25] : Sky marshals may be an appropriate answer to hijacking by the armed hijackers. But four or five fanatic terrorists can overpower a single sky marshal. Moreover, every flight flying more than one sky marshall would be very expensive both for the airlines and for the government. The operators cannot perceive direct benefit of flying sky marshals. Airlines will definitely drop the idea and try to cut cost even if they are forced to fly these men in arms.

Pilot and Cockpit Protection[26] : The doors connecting the cockpit and the passenger cabin must be totally secured as done by the Israelis. It should be ensured that these doors can resist small arms fire and cannot be broken down without considerable force. But in case of sudden decompression there should be ways to release cockpit pressure or else structural damage could take place. There will be certain cost for such modification to existing aircraft cockpit but that cost would be much lesser than flying sky marshals on every flight.

The pilots may be armed with small arms to defend themselves. The pilot need not to carry arms with him in every flight like a gun-toting marshal. The small arm can be hidden in the cockpit in a locker box known to only the pilot and the co-pilot. Proper selection of ammunition could be made before such a step is taken. All the pilots and co-pilots may undergo a minimum training to learn to use these weapons effectively. The secured door should give sufficient time for one of the pilots to gain access to this hidden weapon and be ready for the terrorists to gain access to the cockpit.

It should be made sure that the pilots remain in the cockpit throughout the flight. Cockpit may be redesigned with necessary food and toilets facilities. If it is not feasible to keep the entire flight crew in the cockpit throughout, then at least at least one pilot must remain in cockpit with doors secured, in case the other member has to go to the passenger cabin to attend to some inflight emergency. In the case of hostile takeover or any other medical emergencies the pilot must land the aircraft immediately. A close circuit camera system should be placed to monitor the happenings in the passenger cabin more so near the cockpit door. Any attempt of hijacking or breaking open the door will alert the pilot to land immediately. In the case of a hijacker threatening or actually killing hostages the pilot must land immediately. The local authorities then would deal with the hijackers.

Pilot Response: There must be some standard operation procedure (SOP) for airlines pilots.

- No sooner they come to know about the hijacking they must immediately contact the ground station or the country's air defence center as the case may be.
- Similar to anti-collision beacons, hijacking indicator beacon may be incorporated by the manufacturers of aircraft and the pilot should switch it on along with other electronic indicators.
- 11th September hijacking cases brought out the importance of cell phones. Cockpit crew should carry and contact the intelligence agency in addition to informing the ground station only.

Passenger Response: The passengers of the United Airlines showed rare courage to confront the hijackers and foiled their attempt to crash the aircraft in another sensitive place in the USA. This plan was spontaneous. Must have been through body signals. Yet it was effective. It is a matter of ill luck that the aircraft could not be saved. But such prevention plan can be put into place by proper briefing. Getting more than one passenger to act simultaneously to confront the armed hijackers would require much coordination than is possible. But passengers are responsible to act in certain manner in case of an emergency. A plan to how to resist hijacking can also be a part of the pre-flight briefing in addition to other briefings on aircraft emergencies. Awareness among the passengers will grow and they will understand to act together in case of a crisis situation.

Terrorism is no more a low-tech phenomena. This has become a hi-tech, and very lethal in the hand of well educated and motivated youngmen in their late twenties. This new trend brings out the fact that they are typically part of an obscure network of militants operating transnationally[27] . They do not belong to a organized group with complete chain of hierarchy. They are capable of operating autonomously, helped by the global communication system and Internet. They do not mind engaging in mass murder of civilians and consider this as necessary to strike a blow to their hated enemy. They are driven by religious fanaticism and are young zealots ready to martyr themselves in the course of the attack. Indiscriminate mass terror causing massive casualties by adopting innovative tactics and weaponry are the main area of interest. Ramzi Yousef's[28] WTC bombing in 1993 in New York followed a similar track as discussed here. Following the WTC attack, it was revealed that he had been planning to bomb a dozen of US airliners over the Pacific region.

Probably 11th September attack had been a follow up of those plans. It didn't conform to standard pattern of hijacking but conforms to the emerging trend of international terrorism. The aircraft were never intended to land. These were used as WMD. The intelligence community should expect repetition of such attacks. Spread of globalization, communication, electronic transfer of funds and the growth of air travel will facilitate these desperate groups to operate transnationally. Self destruction and willing to achieve martyrdom by the suicide bombers will complicate the situation further. Positive identification of personal baggage before boarding a flight had been a good deterrence against carrying bombs in the luggage. But these new brand of terrorists in the quest of earning a place in heaven could blow themselves up along with others. Then how to stop these Al-Qaeda type of attackers? The measures at the airport to prevent carrying weapons etc., have to be very strict. Israeli experiences could be emulated all over the aviation world.

Cockpit doors must be secured with necessary modifications. Pilots are to be trained to handle small arms and the passenger cabin should be continuously scanned with close circuit TV. In addition, the frequent fliers will have to learn to resist these criminals from taking over the control of the aircraft. Increased cockpit security and passenger cooperation will definitely make the life more bearable against the future cohorts of Osama bin-laden.

Notes:

1 Jane's All the World Aircraft 2000-2001, p625.
2 Jane's All The World Aircraft 2000-2001, p623.
3 Thomas and Mark Hosenball, "Bush: We're at War," *Newsweek*, September 24,2001 pp 29-30.
4 Ibid p30.
5 Ibid. p28.
6 Wikipaedia:MohammedAtta<www.wikipedia.com/wiki.fegi?action=browse&id=Mohammed _Atta >p1
7 Ibid
8 Ibid
9 Ibid p2.
10 <www.wikipaedia.com/wiki/Ziad_Jarrah>
11 <www.wikipaedia.com/wiki/Khalid_al_Mindhar>
12 Kenneth L. Woodward, A Peaceful Faith, A Fanatic Few, *Newsweek,* September 24, 2001, pp64-65.
13 <www.wikipedia.com/wiki.fcgi>
14 <www.wikipedia.com/wiki.fcgi>
15 <www.wikipedia.com/wiki/World Trade Center
16 America Under Seige, Issue 1,2001 *Multimedia International,* LLC, New York, p64.
17 < www.wikipaedia.com/wiki/Pittsburgh> pp1-4.
18 Ruwantissa I.R. Abeyratine, Aviation Security, Legal and Regulatory Aspects Singapore, Sydney Ashgate, Brokefield USA, 1998, p143-151.
19 Ibid pp156-163.
20 Ibid pp164-173.
21 Ibid pp131-136.
22 Ibid pp171-172.
23 Ibid pp136-138 " Convention on International Civil Aviation, opened for signature at Chicago on 7 December, 1944, entered force on April 4, 1947, ICAO doc, 7300/6.

[24] Ibid. pp14-16.
[25] <http://aspsmith.com> p1
[26] Ibid p2-3.
[27] Paul Wilkinson, "*Emerging Global Aviation Security?*", Paul Wilkinson and Brian M.Jenkins, Aviation Terrorism and Security (London and Portland, Franc Cass,1999) pp151-152.
[28] Ibid. p152.

14

Astrofuel:

An Alternate Energy Source

Currently India's oil consumption is around 84 million tonnes (m.t) per year as against the production of 33 m.t. India's domestic production will touch 37 m.t. by the end of Ninth plan period. On the other hand a sustained GDP growth of 6% will enhance the oil consumption to the tune of 275 m.t. by the year 2020. *The Times of India* wrote on 23 August, 2001, that India is talking to International partners to establish a gas pipeline through Pakistan from Iran despite the present Indo-Pak relations. How India is to break the ice of the present political impasse with an intransigent Pakistan is not known. A long distant hope of generation of goodwill and a desperate need to secure energy source may have driven the Indian policy makers to make such a move. But a comparatively green political scientist would come to a better conclusion. This step is nothing but anxiety driven. The concern for energy security looms large over the Indian psyche. Natural gas reserves of India are placed at 660 billion cubic meters (bcm). India is not an importer of Natural gas at present but it will start soon. By 2002 India is expected to import about 2.2 billion cubic meters which would increase to 25.5 bcm by 2010[1]. Indian desire for a pipeline through Pakistan is well understood.

India is on a better wicket with coal production, which touched a figure of 288.65 million tons by 1996-97. But burning coal has its own limitations. Environmental hazards like acid rain, green house effect and other health hazards will curtail its utilisation.

India also has Uranium reserves, which could generate power to the tune of 10,000 MW. Currently only 2.6 percent of India's power consumption is met by nuclear energy[2].

Indian population has already crossed that magical figure of a billion and is growing at a steady rate. The energy sources on the other hand have been reducing at a much greater pace. The decline is alarming. The reserves beneath mother India would not sustain her teeming billions. Indian electricity generation capacity by 2011-12 should be around 318,000 MW[3]. This is 3.7 times the installed capacity of 85,019 MW at the end of the 8th plan in 1996-97. The projected growth of atomic power is to the tune of 5105 MW in 2011-12 as against the present capacity of 3105 MW (2001-01)[4]. The entire production of atomic power is based on fusion reactors in association with its waste management hazards. A country like India will find it difficult to manage the nuclear waste. While discussing nuclear waste management W Hafele commented that,

> "If the fission products are solidified significantly in a reprocessing plant, then the volume required to store the waste for the next 30 years is only 1,000cubic meters. This calculation assumes that gradually up to 50 to 60% of the energy will be supplied by fission reactors. The waste must be stored under special conditions for five to six years to enable the heat to dissipate. It may then be included in glass or other chemical compound forms which are both stable and sufficiently heat conductive. The most promising long-term method of disposal is in salt mines. One of the major concerns in storing the radioactive waste is that it does not contaminate water either groundwater or ocean."[5]

A Pugwash Monograph (Page 256-257)

The entire picture may not look very healthy. The environmental hazards related to power generation had been commented upon fairly severely in *The Times of India* in its issue of 30 August, 2000. Wherein Proful Bidwai wrote that:

> "..... the distinction of having six of world's 10 dirtiest nuclear reactors. It won't do to deny India's poor safety culture and disaster proneness. It would be disastrous to let hubris drive us toward nuclear catastrophe."[6]

Now the question arises can India be self-sufficient in energy generation? The answer will be in the negative. Planning Commission has accepted this reality and placed on record their estimated gap of demand and supply. Improving energy efficiencies is costly and will not keep up with the energy demand of India.

> " Environmental problems associated with burning coal will put downward pressure on coal consumption. Long lead time, political sensitivities on the international stage and difficulties in attracting the necessary foreign investment will complicate India's ambitious plans to expand the nuclear power sector."[7]
>
> Juli A Macdonald and S Enders Wimbush,
> Strategic Analysis (August 1999)

Not Too Bright a Picture for Asia

Oil demand growth in Asia may be the fastest than in any other region. It is expected that demand this year (2002) will be around 13.3 million barrels a day (mb/d) almost touching the West Europe's 14.3 mb/d. Asian demand will rise to 28.6 mb/d which is greater than total US consumption of 24.4 mb/d. By 2020, China's oil consumption is expected to be 9.5 mb/d.[8]

Demand and supply of natural gas in Asia grow at a much faster rate and will take the seat of primary source of energy over the next two decades. This consumption is expected to grow by more than 7% till 2020. From 4.7 trillion cubic feet in

1995 it will rise to 27.7 trillion cubic feet. In comparison, consumption of gas will grow at a much slower pace in Australia, Japan and New Zealand at 1.6 percent a year (from 3.1 trillion cubic feet will go upto 4.6 trillion cubic feet by 2020)[9]. This would be the preferred fuel in Asia because of its availability, cost stability and environment friendliness. Natural gas could reduce the dependence of Asian countries on Persian Gulf supplies.

Coal consumption in Asia is the largest in the whole world and that at only 40% level of the world's consumption. And it is expected that Asia's share of coal consumption will rise to a level of about 60% by the year 2020. In terms of tonnages this will be about 5.2 billion tons. Asia should brace itself for the associated environmental problems due to such heavy burning of coal.

It is also expected that nuclear power growth in Asia would rise from 85.5 to 129.4 GW by 2020[10]. But not without its corresponding waste management hazards. All such statistics only point out to Asia's dependence on energy from external source. The strategic reality is clear as commented by Juli Mcdonald and Wimbush[11]. The Middle East will become more important. Or should it be?

World Energy Resources: Presently the world's average annual energy consumption is 71,530 barrels per day as against the total world reserve of about 1,034,668,270 thousands barrels[12]. Therefore a conservative estimate indicates that the world fossil fuel will be exhausted in 39 years from to date. And when the population reaches 10 billion mark the consumption will be to the tune of about 100 to 150 billion barrels of oil per year. The economically recoverable fuels left on Earth are equivalent in energy content to only 5000-7500 billion barrels[13]. And by the midst of this century, fossil fuel will not be in a position to sustain human civilization. Mankind will have to look for new source of energy based on solar wind, hydro, geothermal and biomass.

Nuclear fusion is the other source of energy but for its present problem of production of greater number of neutrons, which damage the reactor itself and produce radioactive wastes. On the other hand, a fusion reaction carried through Helium 3 releases only 1 percent of its energy in the form of neutrons. As a result the design factor of this type of reactors become easy and reduces radio activity to a very low level. This may be mentioned here that **Astrofuel** would be the most efficient source of energy because 99% of the energy can be released as charged particles. This energy can be converted into electricity with greater amount of efficiency. The level of radio activity is so low that a complete meltdown of the reactor will not spread the much dreaded radio active particles[14]. Or at the end of the useful life of the reactor, this could be dismantled and disposed like any other ordinary medical instruments.

An Alternate Source

For the first time in 1986 the Wisconsin Centre for Space Automation and Robotics at the University of Wisconsin-Madison conceived the idea of mining and using Astrofuel from the moon[15]. This centre coined the word **Astrofuel**. The center, one of the 16 NASA centers for the Commercial Development of Space, is well positioned to manage this project because of the university's already existing fusion, space and life support research programme. Researchers here do not have any doubt that 21st century power plants would wear different look from their predecessors of the yesteryears. They are more or less sure that Helium3 (He3) an isotope of Helium which has one less neutron than Helium, available in plenty on the surface of the moon could replace the fossil fuel. The world population touching a figure of almost 10 billion would find it impossible to substitute the energy need from any source other than Helium3. This is a rare quantity on earth but available in large quantity on moon[16]. One ton of this Astrofuel could supply the electrical needs of a city of 10 million people when combined in a fusion reactor with a form of hydrogen extracted from water. Therefore it won't be very difficult to understand the impact Astrofuel can have on world energy supplies. The extremely high power density means that only 28 tons of Astrofuel approximately the payload of current U.S Space Shuttle could supply the entire electrical demand of the US for a year. Even if the selling price is set for $ 1 billion a ton, the energy cost of this fuel would be equivalent to oil at only $7 a barrel (University estimates)[17].

In the early 1970 for the first time the astronauts discovered quantities of Helium3 on the lunar surface[18]. It originated from solar wind and got deposited over thousands of years. Therefore moon could be the most important source of energy in the 21st century. Vastly more energy can be supplied by Astrofuel than by oil from the Persian Gulf. The nation that develops the technology to retrieve Astrofuel will be in a commanding economic and strategic position in this century. Astrofuel will provide the basis for:

- Economic and inherent safe fusion reactors to supply electricity for the world.
- A nuclear fuel cycle with radioactive isotopes.
- Energy converted directly to electricity with much greater efficiency.
- Completely and inherently safe reactor operation. In the event of the worst accident no meltdown could take place.
- Radioactivity of the reactor is so low that the facility can be disassembled at the end of its life and disposed off as low level waste.
- This is also interesting to note that human existence on moon and in space can be sustained by the by-products of Astrofuel. The process of mining on the moon will produce certain quantities of[19]:
    - o Nitrogen
    - o Methane
    - o Helium
    - o Water
    - o Carbon-Oxygen compounds
    - o Hydrogen 6100

These elements and compounds are vital to human existence in space. They will permit growth of food, establishment of contained atmosphere and water supply. Indian scientific community and the energy researchers should take note of these developments. We may not have much of a choice than to accept the fact that the US now has the overall lead.

Helium3. As has been mentioned earlier, Helium3 is an isotope of Helium, which has one neutron less than Helium.[20] Some quantity of He3 is available on earth. But this quantity is not sufficient to be exploitable commercially. The total US strategic reserve is not greater than 29 kgs and another 187 kgs are mixed up with the natural gas. The moon has an estimated reserve of 1,100,000 metric tonnes of He3 deposited by the solar wind for millions of years.

Table I

Specifications of Helium3

Chemical Symbol	He3
DOT Classification	7440-59-7
Substance Identification	1046
UN Number (Compressed Gas)	UN 1046, Class 2.2
Major Hazards	Nonflammable gas
Toxicity (TLV)	None Established
Flammability Range (in air)	None
Odor	None

Data from Rhombic Corporation–Helium3 (www.rhom-bic.com/helium3.html)

Table 2

Producers Purity Specification of Helium 3

Typical Contaminants		Spectra Grade Chemical Purity ** 99.999% Isotopic Enrichment * 99.99%	Refrigeration Grade* 99.9%
Argon	Ar	0.5 ppm	1.0 ppm
Carbon Dioxide	CO2	0.5 ppm	0.5 ppm
Helium-4	He	100.0 ppm	1000 ppm
Nitrogen	N2	0.5 ppm	0.5 ppm
Oxygen	O2	1.0 ppm	1.0 ppm
Water	H2O	0.1 ppm	0.1 ppm
* Maximum Contaminant level			
* Chemical Purity is exclusive of Helium-^{4} impurities			

Data from Rhombic Corporation Helium 3 (www.rhombic.com/ helium3.html)

Energy from He3: One million metric tons of He3 reacted with deuterium would generate about 20,000 terre-years of thermal energy[21]. A terrewatt- year is one trillion (10^{12}) watt-years. One 100 watt bulb uses 100 watt-years of energy in a year. That would indicate what a terre watt year mean. Even if we use all the uranium available on earth and use it for energy production we could generate about half this energy and face some disastrous situation safeguarding the nuclear wastes. And in terms of fossil fuel the level of energy will be 10 times more than the volume we could get from the fossil fuel.

It has been determined that the sea of tranquility is the prime target for mining of **Astrofuel** because many tons of He3 are deposited below the surface[22]. The other most important area for mining could be close to Mare Serenatatis. This was discovered by the Apollo 17. Models to explore moon are yet to be built but it has been more or less determined that solar powered robotic equipment would scoop up the top layer of lunar soil, place it into a rcbotic unit. The soil would be heated upto 600^0 C thus separating the He3 from other lunar material[23]. The spent material then would be dropped off the back of the moving robotic miner. Because of moon's gravity being one sixth that of earth, relatively little energy would be required to lift the material. No tracks would be left by the miner because the soil would be smoothened and slightly fluffed up as the spent material redeposited on the surface[24].

Commercial viability of the Astrofuel was determined by the University Research Center in the year 1987. At 1987 cost, it was found that the US spent \$ 40 billion to buy coal, oil, natural gas and uranium to produce electricity. For supply of similar Mega Watt of electricity for one year the US will import only one shipload of fuel and may spend 25 billion dollars. But in terms of dollar to barrel ratio it will work out to be only \$ 7 a barr[illegible]dicated earlier.

Astrof[illegible] is not merely a subject for scientific curiosity, it will have a major impact on industries of the US and other participating countries. Billions of dollars will be required for research and development. Implementation of use of Astrofuel on earth will require many technologies to be created. Foremost among them are super conducting magnets, plasma control and diagnostics, robotically controlled mining equipment, life support facilities, rocket launch vehicle, telecommunications, power electronics, etc. Though the investment seems to be astronomical but if compared to the benefit derived, the justification will be more than adequate. For example, the amount of coal required to generate 1,000 MW of electricity will be the effort equivalent of half a gallon of Astrofuel per day. And since fusion plants emit no ash or particles, electrostatic precipitation equipment would be unnecessary. The US industries have got a head start to use the Persian Gulf of the 21st century to make it a reality. The commercial viability is enormous. Apart from the US requirement it is estimated that an additional quantity of 75 to 100 tons of the fuel would be required for the rest of the world. At an average cost of 1 billion dollar a ton, the US industries should be able to scoop up an equivalent whooping sum and at that point of time when the fossil fuel will be really scarce i.e by the late 2030s. By that time the world will be energy hungry in real terms.

Americans view the 20th century as the century of the Americans. Triumph of Western liberalism over communism had brought about an ultimate stability to the world and the new world order that has emerged may be not fighting another world war but fighting economic wars. Jasjit Singh while commenting on this, stated that " The bipolar world formally ended with the demise of the Cold War, although it was showing signs of change for quite a many years before that. But the evolving international order is neither unipolar nor multipolar. The former assumes that the solitary "pole" (in this case, the United States) possesses the requisite capabilities and is willing to exercise control and influence over the remaining 183 states in the world."[25] Astrofuel source, totally controlled by the Americans, will be the ultimate weapon they would desire to have.

The debate in the USA about a new world order was actually mooted by President George Bush in September 1990 just before the Gulf war. But he never went on to explain what he meant by the new world order. However it seems that the new world order was nothing but an order seen through an American prism. The US Air Force motto, "Global Reach Global Power" may be an indicator of the US view of the 21st century. President Bush's interventionist overture in the Gulf crisis to traditional allies was accepted well and they responded enthusiastically. Who were those allies? And why did they readily accept the American invitation to join hands to throw out Saddam's troops out of Kuwait? It was nothing but the instinct of survival and maintain the economic equilibrium which hung most precariously on the barrels of oil produced from the field of Arabian peninsula. Subjugation to Saddam's game plan and accepting such daylight banditry could have spoilt the Western imagination of the post-communism world order. The USA led the Gulf war to its successful completion. A United States who was not ready to commit its ground troops subsequent to the Somalian experience committed everything possible in its inventory to throw Iraqi's out of Kuwaity soil. What could have been the driving force for George Bush to behave in a fashion and make the US people understand the urgency of such an action? The US population never forgot the year 1973 when they had to queue up at the petrol pumps to fill up their car petrol tanks. They could not accept such situation again. George Bush only had to remind his fellow countrymen about 1973 crisis and Americans rallied behind him. They wanted the scalp of Saddam Hussain. Source of oil couldn't be disrupted. It was adequately demonstrated that the world order is driven by self interest. Americans understand this more than anyone else. "Global Reach Global Power" option is exercised even to a distant threat like Osama bin-Laden. Some years back when Buchanan[26], a syndicate columnist, decided to stand for the Presidential election he attacked the US foreign aid programme and brought in the isolationist theory. The US was not required to maintain its global presence, as the Soviet threat no longer existed. There were others in the USA who joined the chorus. With such developments on the mainland of the main global actor, it is really not understood what will happen when the principle source of energy, the fossil fuel, will run dry and be non-existent or will be available at not an affordable price. The present oil price level should not put the countries in a state of complacency. Everything that is good will be over by the year 2030. The belts will have to be tightened by the year 2020. The USA is on the lead for the exploration process of a new source of energy and they are almost in the process of a major breakthrough. They have already declared that moon will be the Persian Gulf of the present century. One, half a gallon can of Helium3 will do the work of more than 1000 tons of coal. And who will own this real estate? The only source of affordable energy source will be completely dominated by American industries. What will happen to the new world order as envisaged by George Bush or a Ultra New Order will emerge completely dominated by America. India along with other United Nations members will stand in a queue to receive the quota of fuel fixed by "Uncle Sam". All discussion about energy security will take back seat. The United Nations will discuss everything but the nations will wait for their quota to arrive. Astrofuel will decide politics, economics and world order.

The technology to harness He3 energy source is an on-going process in the laboratories of the US Universities. Miniaturization of the He3 driven reactors would have immense dimension in the future ultra new world order. The Kursk tragedy will be forgotten, never to be repeated by an errant torpedo. This safe energy could propel submarines ships and airships for unlimited period. Dimension of such fantasy may overtake Jules Verne but could be transformed into reality.

Indian self-sufficiency in the energy sector is a distant dream. Planning Commission in their report has very clearly established the gap between the demand and supply but could not give a satisfactory answer how this huge gap could be bridged. Although oil and gas supplies are in abundance in world but what could happen 15 years hence is anybody's guess. India will have to form alliance for energy security. Will India be in a position to engage Pakistan in an useful dialogue in an area of mutual concern to secure energy source and have greater access to Central Asia. Complication of India's energy security interest will arise and not too at a distant future because the US will gradually disengage herself from the Middle East and look elsewhere for its energy requirement. Could Astrofuel be one of them? Indian dependence on energy source will increase rapidly and it would be better for Indian scientists and energy security analysts to analyse this unconventional source and secure a place for India before it is too late.

Time is ripe enough for India to take up the Moon Treaty[27]. The Treaty of 18 December, 1979, is collecting dust over last twenty years and it has to be debated upon by all the UN members and to be ratified. India will find many willing participants if it decides to take a lead. Article 11[28] of the treaty adequately covers the interest of the entire world community when it states that " The moon and its natural resources are the common heritage of mankind, which finds its expression in the provisions of this agreement. And the moon is not subject to national appropriation by any claim of sovereignty, by use or occupation, or by other means." Americans won't be amused but it won't be difficult to convince the world body.

Notes:

1 Shebonti RayDadwal "Energy Security: India's Options" (*Strategic Analysis* July1999).pp 653-655.

2 Ibid.

3 Ninth Five Year Plan (GOI Official Publication) p106.

4 Ibid.

5 "Impact of the New Technologies on the Arms race" (*A Pugwash Monograph*, Library of Congress, Catalog No 72-148970).

6 Proful Bidwai, "Sunken Hopes: Lessons from Nuclear Sub Tragedy" *The Times of India* 30 August, 2000.

7 Juli A.Macdonald and S.Enders Wimbush "India's Energy Security", (*Strategic Analysis,* August 1999.) pp821-835.

8 Ibid.

9 Ibid.

10 n7 p823.

11 Ibid.

12 *Energy Statistics Source Book* 14th Edition.(Tulsa, OK 74121: Penwell) August 1999.

13 Oil and Gas Journal Energy Database publication (Tulsa, OK 74121).

14 Lunar Mining Could Provide Future Energy Source for Earth (*NASA News Release 90-10-09*. 10/70/90. Release 90-139).

15 Ibid.

16 Ibid.

17 "Astrofuel for the 21st Century" (Handout by the College of Engineering, University of Wisconsin-Madison) 1993.

18 Harrison H Schmitt "The Real World and Lunar Base Activation Scenarios" (NASA Conference Publication 3166, Vol2 pp 667-671).

19 NASA News Release 90-10-09.

20 "Helium3" (www. rhombic.com/helium3.html/)

21 Ibid.

22 Harrison H Schmitt. "The Real World and Lunar Base Activation Scenarios" (NASA Conference Publication 3166, Vol2) pp 667-671.

[23] Ibid.

[24] Mining in Manitoba, (http://11204.112.230.2/minsci/future/moontreaty.htm)

[25] Jasjit Singh" India's Strategic and Security Interests" Discussion Paper for the Sixth IDSA–INSS Strategic Symposium, New Delhi January 24-25, 1996.

[26] Chintamani Mahapatra "Indo-US Relations: Into the 21st Century" (New Delhi: *Knowledge World*), p143.

[27] "Moon Treaty" (www.iasl.mcgill.ca/space/moon.htm.)

[28] Ibid.

15

Indian Nuclear Dilemma

This is definite that India cannot be credited as the inventor of the strategic option of "no first use " nuclear strategy. In 1982, four former senior US officials—McGeorge Bundy, George F Kennan, Robert McNamara and Gerard Smith—tried to force a resolution proposing that NATO adopt a "no first use pledge[1] ." In this pledge NATO would promise not to initiate the use of nuclear weapons even if Western Europe was being overrun by a Soviet conventional force. This was quite a departure from the Mutually Assured Destruction (MAD) theory as propounded by the same officials earlier in their life. Because "no first use" principle repudiates" extended deterrence". As a result Indian declaration of" no first use," came as no surprise to the world. Because such nuclear strategy was already known and had been discussed for many years.

The "no first use" policy declared by India reinforces the defensive nature of India's nuclear posture. And it makes clear to the world community that the nuclear forces will not be structured to initiate war fighting despite building up sufficient capability assisted by its missile programme. These would be poised for retaliation to cause unacceptable damage to the adversary. This defensive posture implies the deployment of a small but sufficient nuclear arsenal to guarantee the second strike capability.

No First Use, Massive Retaliation, Mutually Assured Destruction (MAD) etc., and the other nuclear strategies which were the intellectual property rights of the strategic intellectuals had almost become a reality. The subcontinent could have become the testing ground of all these when India and Pakistan began saber rattling to fight the next round of war in the summer of 2002. Pakistani President General Musharraf made no bones about using nuclear weapons if attacked by India. In every forum he declared to teach India a lesson. Even the UN representative of Pakistan Munir Akram declared openly that Pakistan would not hesitate to use its nuclear arsenal if attacked by India. Defensive nature of Indian nuclear response was put to severe test. The theory of deterrence was put to severe test. To provide an answer Col (Retd,) Bokhari stated that, "Deterrence is a very delicate issue and a subtle one too, perhaps too subtle for easy human perception. In the absence of experience of a full-fledged nuclear war the issue becomes still more evasive and intangible. It may vary from country to country[2] ." Western countries took no chances; and they asked their citizens to leave the subcontinent.

Indian Nuclear Programme

In May 1974 for the first time India demonstrated its nuclear capability which was declared as Peaceful Nuclear Explosion (PNE) by the then Indian Prime Minister Indira Gandhi. But the programme was not carried forward to a step further to install Indian nuclear bombs. Indian self imposed moratorium on nuclear agenda was taken for granted and no credit was given for such self restraint. India didn't have any choice but to refuse to subscribe to the Comprehensive Test Ban Treaty (CTBT). Because India found to its dismay that not only the treaty was discriminatory, but more importantly, it contained no potential to further the cause of disarmament.

Indian security environment on the other hand continued to deteriorate as a result of nuclear and missile proliferation. There were continuous reports from international intelligence agencies that Pakistan had already got the nuclear capability. Added by the missile capability given to it by China and North Korea[3] . Both these nations have dismal respect for Missile Control Technical Regime (MCTR), CTBT and other Western generated regimes. Western nationals did not have any tool to punish these two erring nations. Pakistan utilised this ambiguous situation to its fullest extent and went ahead with its clandestine nuclear bomb making effort and missile development. As a result India found itself increasingly and involuntarily embedded in a heightened nuclear environment.

India in Post Cold War Period

Nuclear disarmament as talked by the super powers in the cold war period emanated from the fear of nuclear holocaust and realpolitik of the period dictated the pursuit of nuclear deterrence. No doubt the theory did work. There were no wars after the Second World War when the bipolar world could have gone to war with each other. Of course there were times of anxiety like in the case of Cuban missile crisis. The real danger of a no win situation was very clear to the two principal actors. They decided to sign the Anti Ballistic Missile treaty, and began a dialogue for the START I and START II etc. Unfortunately, the end of cold war has not led to any significant reduction of nuclear weapons and dependence on deterrence; paradoxically, nuclear deterrence has gained importance for nuclear powers, with its role enhanced to cater for an era of strategic uncertainty.

This followed by legitimization of the unequal nuclear order in perpetuity, through the extension of NPT and orchestration of nuclear arms control treaties like CTBT and FMCT to the absolute advantage of the Nuclear Weapon States (NWS). India was pushed into a corner with its options gradually being reduced to almost zero. These challenges posed by the geostrategic environment in the last decade, made it necessary for India to opt for the tested strategy of " Nuclear Deterrence," after having exercised self imposed restraint form 1974. Indian sincerity of total nuclear disarmament was not taken seriously by the NWS.

They continued in their effort towards greater sophistication of the missiles and the nuclear devices.

China joined the nuclear race and desired to replace Russia as one of the principal actors. It went ahead to assist Pakistan to gain access to this technology despite the US protests. North Korea delivered the delivery vehicles. Pakistan could arm itself with the missiles like Ghauri, Saheen, etc. All of them are Indo-centric. Because Pakistan does not consider the other Islamic nations bordering Pakistan as its enemy. Moreover, Pakistan gradually losing the conventional arms race had to depend on the nuclear weapons to carry out its misadventures like cross-border terrorism with impunity.

India could not watch this entire drama helplessly. The situation gradually becoming hopelessly anti-Indian. Therefore, India was not left with any option but to take necessary steps to ensure that the nuclear option, developed and safeguarded over decades, was not permitted to be eroded by a voluntary, self-imposed restraint. Guided by the concerns of national security and in continuation of it's policies of self-reliance, India conducted the nuclear tests on 11 and 13 May 1998, thereby emerging as a Nuclear Weapon State. Pakistan followed soon with its own tests which vindicated Indian stand about Pakistani nuclear bombs. India became ready to accept the condemnation of the Western world and sanctions of the USA. Nuclear muscle of Pakistan gave it the legitimacy of escalating tension at the Kashmir borders. Despite Prime Minister Vajpayee's singular effort to extend the olive branch at Lahore, Nawaz Shariff and his Army Commander in Chief, General Musharraf, backed by this new found power, decided to cut off India at Kargil from its extended Northern region.

Indian Nuclear Tests

Again it was Indian Prime Minister who announced to the entire nation about this singular achievement of the Indian scientists and called it as " Jai Vigyan." Shakti tests clearly demonstrated that India has the requisite level of expertise in modeling and simulation. The yields were as predicted which indicated the broad based weaponisation capability of the nation. India carried out a total of five nuclear tests, which were thought to be sufficient for computer simulation of new designs and would also permit advances in sub-critical technology in the future as claimed by the Indian scientists. But the Western nuclear scientists expressed their reservations. At the same time India reiterated its desire to exercise stringent controls on the transfer of sensitive nuclear technologies and equipment.

Nuclear Environment

India is a part of the Southern Asian region, but areas stretching from West Asia through Central and South Asia to South East Asia and East Asia are regions of considerable strategic interest to us. The region includes declared nuclear weapons states and nuclear/missile capable states, stretching from North Korea, China, India, Pakistan, Iran, Iraq and Israel as strategic components of consequences. It is also well known that our Northern neighbour is a declared and accepted Nuclear Weapons State (NWS) and it was known to Indian strategic community that Pakistan had achieved that status, albeit covertly earlier, and overtly in 1998. Thus India has been embedded in a nuclear environment for long.

China maintains a full range of nuclear capability of about 400 warheads[4] . Around 250 strategic weapons are in a triad of long range land based missiles, bombers and submarine based ballistic missiles. Out of these it is expected that the bombers will be phased out and China will bank upon the recently inducted SU 27 or SU 30s to carry the nuclear weapons if at all. Chinese SLBM programme has not been a success story but considering Chinese ingenuity, it would soon be able to install this important leg of the nuclear triad. It also reportedly has 150 tactical weapons presumably with lower yield meant to be dropped by tactical aircraft, artillery shells, atomic demolition munitions and short range missile. China with its ' Limited Nuclear Deterrence," doctrine and " No first use" policy, except in extreme national interest within areas it claims as it own, enjoys a clear asymmetry in the nuclear field and would remain a potential threat.

Whereas Pakistan's nuclear arsenal is India specific. During the recent crisis of the summer of 2002 this was adequately articulated by the present Pakistani president General Musharraf, who has only one point programme i.e "hate India and threaten India," by brandishing the nuclear weapons. With such propensity towards brinkmanship and rejection of " No first Use," principle, Pakistan thus poses a direct and substantial nuclear threat even with its current capabilities.

Indian experience with China and Pakistan has to be taken into consideration for formulation of all future policies whether conventional or nuclear. China has had a close defence relationship with Pakistan for many years. It has supplied M 11 missiles, which can only be targeted against India. There is clear evidence of Chinese collusion in building the nuclear capability of Pakistan, starting from transfer of the basic technology in the late 70s, to the supply of ring magnets to Pakistan. This collusion is well documented. There are strategic overtones to this relationship, in which China appears to be using Pakistan to keep India engaged militarily.

In such intense nuclear environment it would be important to understand and analyse the threshold of a nuclear war, which is different for each one of them.

- Indian Nuclear Response:
  - o Indian threshold of nuclear response derived from known position of the National Command Authority. And as such it would be used as a positive retaliatory response to a first strike by an adversary.
  - o Also may be a response to the use of Tactical Nuclear Weapon (TNW) by Pakistan on our troop concentration. Discounting the proximity of the deployment of own troops.
- Pakistan understands very well that it can never win a conventional war against India and the asymmetry is ever growing with its dwindling resources and economic

difficulties. That is why Pakistan's compulsion to use nuclear weapons could arise out of the following:

- o When loss of a vital area is imminent like Sind, large areas of POK or a major city.
- o Near exhaustion of its conventional war waging capacity.
- o Imminent collapse of the state consequent to severe military reverses.

- Chinese threshold for using nuclear weapon would be totally different. Because China no doubt enjoys a total superiority of nuclear and conventional forces over India. It could use these to coerce India at any point of time and that is why Chinese threshold would depend upon the following:
  - o Indian forces gaining substantial advantage across the border resulting in loss of face to the Chinese.
  - o Emergence of a threat to China lifeline such as its Aksai Chin highway and other communication routes to Tibet.

Nuclear Objectives

Like any other nuclear forces, Indian nuclear force is to deter the use of nuclear weapons against India, as also to dissuade any form of nuclear coercion or blackmail. Deterrence being the core value of any nuclear force. Here in India also these would also reduce the risk of use of chemical and biological weapons against India and serve as a counterpoise against an overwhelming conventional threat in the event of supreme national interests being threatened. On the other hand a credible nuclear force should be in a position to threaten the enemy targets which it considers important. And should deterrence fail, the forces should be accordingly so structured, deployed so as to provide options to the national leadership. The addition of a nuclear force has not in any way brought any change in force structure of conventional forces, nor is it likely to. The mere fact that the nuclear weapons cannot be used for war fighting and the self-imposed stipulation of No First Use (NFU) demand that the conventional capability cannot be reduced in any substantial manner, unless the immediate adversaries of India show an inclination.

Nuclear capability had always been the stepping stone for international recognition and acceptability in the comity of nations. That might have been the reasons for the smaller nations like Iraq, Iran etc., seek power through nuclear route. And it is certain that the US couldn't have attacked Iraq or could carry out Operation Desert Storm if Iraq had nuclear weapons in its possession. And this nuclear capability displayed by India would enhance Indian strategic space in the international affairs by providing legitimate leverage in international negotiations.

A credible nuclear force must be diverse, flexible, effective, survivable and responsive. The NFU clearly underlines the nature or degree of survivability. None can afford to give such undertaking unless sure about own capability and survivability. But nuclear uncertainty that perpetuates from nuclear terrorism and irrationality cannot be totally discounted, where all theories come to a stand-still. For an example, "The tenor and language of the highly irresponsible statements recently emanating from Musharraf and his senior colleagues and officials such as Munir Akram, Pakistan's permanent representative to the UN, remind one disturbingly of the tenor and language of the irrational statement emanating from time to time from Bin-Laden and the jihadi organizations. They show that nuclear irrationality is not confined only to Bin-Laden and his brand of international terrorists, but is also found in the mindset of Musharraf and other leaders of Pakistan's military intelligence establishments[5]."

To measure the effectiveness of any nuclear force and for it to have any credibility it must possess the following inherent qualities:

- **Response Time**: Expected launch time within which a retaliatory strike must be launched is an important determinant of the nuclear forces. Whether the response need be immediate or deliberate with delay would eventually decide the type of structure that would need to be put in place. Structure for nuclear warfare must be in a position for immediate response. But as far as India is concerned, technological shortfall would be the only source of delay, and response time will be governed by capability rather than anything else. India therefore should try to create achievable response time, which would reduce as the technological development progresses further.

- **Weaponisation:** Weaponisation could be considered to be a status wherein the nuclear device is available in the form of warheads with requisite launch vehicles that can be activated and used when so desired. Weaponisation could be in the form of separated warheads, where the warhead and the launch vehicle are stored separately or mated with warheads, where delivery system with the warhead is ready for use. In our context, the former option would be suffice, since India is looking only towards a response capability to act as deterrent. This option has the inherent advantage of keeping the nuclear threshold in the subcontinent at a recognizable level, despite Pakistani utterances.

  If the capability, as pronounced, is that of a minimum deterrence, the emphasis is on sufficiency. In today's thinking, no country is ready to accept even a single nuclear explosion in any of its major cities. Consequently, sufficiency can even be the assurance of one single attack on a major city. This would require multiple deployments, as well as survivability of the vehicle in the emerging future defensive environment.

- **Survivability:** Indian nuclear forces must have the ability to survive enemy attacks to provide the retaliatory strike. Survivability is in fact the key to the NFU policy and will be ensured. And it is ensured by having alternate means of attack in the form of a variety of platforms, sites and locations from all the three elements i.e surface, air and sea. Underwater launch capability would provide the intimate diversity of platforms and more assured retaliatory capability.

The adversary must be convinced that any attack on India will leave sufficient retaliatory capacity to inflict unacceptable damage as a response under any and every such circumstance. Potential aggressor must be convinced that in any scenario, sufficient capability will remain to deliver a devastating retaliatory strike. Survivability is enhanced by a combination of multiple redundant systems, mobility, dispersed and hardened sites, number and variety of launch platforms and mobile employment concepts.

- **Command and Control:** Command and Control structure of any nuclear capable state should be secure and survivable and there should be sufficient redundancy. It should enable the leadership to execute, direct, assess and terminate nuclear options whenever needed. Degradation of the same will be the immediate aim of an adversary, which dictates the necessity for the structure to be robust, redundant and rapidly recoverable. Also it must be hardened enough to withstand the impact of EMP.

Nuclear Policy

Indian nuclear policy can be derived from various statements made in the Parliament by the political authorities. The basic structure of the same may be as such:

- **Minimum Credible Deterrence:** There is no historical experience of bilateral nuclear war and that prevents us from guessing how nuclear weapons should be deployed or how many would be enough to deter an enemy from carrying out any mischief. But it is difficult to prevent anyone from carrying out the first strike[6]. Deterrence is based on the ability to absorb the impact of the first strike and capability to carry out retaliatory or the second strike[7]. There are other theories of deterrence like:

- **Mutual Assured Destruction**: The Mutual Assured Destruction (MAD) theory is based upon the idea that war can be prevented if both sides have a secure second strike capability[8]."
- **Flexible Response**[9]: McNamara first proposed the strategy of " flexible response." The core of this policy was a plan to give NATO a "credible" conventional capability while at the same time threatening to escalate to large scale nuclear use if NATO begins to lose conventional warfare.
- **Massive Retaliation**[10]: " John Foster Dulles outlined a change of direction for the strategic doctrines that had been developing under Truman. This new doctrine, known as one of massive retaliation, was widely assumed to be founded on an undiscriminating threat to respond to any communist inspired aggression, ...by means of a massive nuclear strike against the centers of the Soviet Union and China."

There have been many discussions on the above strategies as had been propounded from time to time but basic purpose of the Indian nuclear force should be to deter the use of nuclear weapons against India and to prevent any form of nuclear blackmail. In Indian context, to define minimum deterrence would be complicated because of the dynamic nature of global power structure. India faces threat from more than a single adversary having different levels of asymmetry not only in nuclear force but in overall strategic capabilities inclusive of military and economic powers. Moreover, nuclear weapons have never been used and they derive their power not from the numbers or yield but from psychological perception and impact on the adversary. And the key issue being how many weapons are deemed to be sufficient. It is very sure that minimum credible deterrence cannot be strictly quantified in terms of number, etc. This is a policy determined in the context of security environment. The size, both in terms of numbers and yield, of the nuclear force would be limited by what is considered to be sufficient to meet the credibility of the deterrence posture vis a vis India's adversaries.

As no country is ready to accept even a single nuclear city busting explosion in any of its major cities or in any part of its territory, sufficiency can only be based on assurance of even one strike breaking through the defences of the enemy. But this should be made sure through multiple deployments as well as methods to ensure survivability of the launch platforms.

Deterrence is based on actual force capabilities and the national will to use these forces if necessary. National will or resolve is projected through policy statements by the leadership. For deterrence to succeed, there should also be credible plans for action in case deterrence fails. And the credibility of the force is determined by its capability and the structure in place to ensure its application.

An adversary must understand our invulnerability and assured reactive ability even to carry out a first strike if required. That dictates our level of capabilities, which should be minimum commensurate with possessing the ability to absorb first strike and then react. Our limited resources therefore, restricts the capabilities of our forces to be maintained at a minimum acceptable level, which must ensure a credible deterrence.

No First Use: NATO was urged to propagate this " No First Use" policy, as discussed at the very beginning of this chapter. Indian Prime Minister articulated categorically about this policy to establish India's defensive nature of the nuclear forces. The nuclear forces will not be structured for war fighting but as deterrence, to carry out retaliation to cause unacceptable damage to the adversary. Such defensive posture implies maintenance of a small but survivable nuclear arsenal to guarantee an assurance of retaliation.

Nuclear Testing: India may be the only example among the comity of nations which put restriction on itself for more than two decades to not to carry out nuclear tests or go on the path of nuclear weaponisation from the summer of 1974. Now it is India again, once satisfied with its 5 tests at Pokhran, which declared a self-imposed and unilateral moratorium on further nuclear tests and has indicated a willingness towards a "dejure' formulation of this declaration as also to enter into constructive negotiations on a Fissile Material Cut off Treaty.

Global Disarmament: India is a Nuclear Weapons State, but that does not take away its commitment towards global nuclear disarmament.

Nuclear Weapon, When?

Decision to use nuclear weapon is entirely with the Prime Minister. And that would be determined by the strategic situation, a significant escalation from conventional warfare or by some action of enemy considered as serious threat to the National security. Based on NFU policy, it is of critical importance that the nuclear forces are capable of adopting heightened states of alert, which would be guided by the following operative factors:

- **Readiness and Alert Status:** Readiness and alert status will depend upon strategic situation prevailing at that time. The NFU will prevent us to carry out first strike even under imminent threat. But that would not inhibit India from carrying out retaliatory strike. The NFU policy must have means of projecting consequences of adversarial actions in an environment of escalating tension. Raising the readiness of nuclear forces can send the desired signal to the adversary. The state of readiness of nuclear forces will therefore be announced so as to be understood by the international community also. Sometimes, an increased readiness levels may be necessary to deter aggression. However, an increased risk of attack prompted by rising state of readiness of the adversary may require that Indian nuclear forces be maintained at visibly increased state of alert for prolonged periods. It must be noted that increased readiness is intended to signal national resolve, but must be accompanied by measures that would allow for de-escalation.

- **Threshold of Use**: NFU as declared and deterrence will determine the threshold of nuclearisation of a conflict. Primary use would be to carry out a retaliatory strike. The use of Tactical Nuclear Weapons (TNW) by the enemy shall elicit a counter value response, and the conviction of this response is the essence of the deterrence.
- **Targeting:** Strategic counter value targets are major cities, centres of high value economic or military targets, where accuracy is not an important requirement. Actual requirement is the deterrence which when applied will cause unacceptable damage to the aggressor.

Dilemma

Summer of 2002 almost saw the next round of war between India and Pakistan. Pakistan armed with its nuclear weapons and the delivery vehicles like Ghauri, Shaheen and Ghaznavi sent clear messages to India and the world community that Pakistan will never accept a defeat or another dismemberment by the Indian superior conventional forces. Such a small war almost became a reality as Pakistan could not cap the transborder terrorism effectively or didn't wish to do it. And another sensitive attack by these militants would have seen the Indian forces pushing into Pakistan. But Indian responses were not very clear. India spoke about limited war under nuclear circumstance. But none could exactly articulate what was the meaning behind it. India might have hoped to carry out limited operations by bombing, shelling or special forces raid on militant camps inside Pakistan controlled Kashmir or even within Pakistan border. Seizing a portion of Pakistan territory could have been a distinct possibility. But as long as the fighting was limited to Kashmir there was little possibility of escalation. The terrain would have acted as the safety net. But if the war would have escalated to the plains of Punjab or Rajasthan, containment would have been impossible. This is where the nuclear weapons come in. Pakistan has built its nuclear forces as India specific because it doesn't have any other immediate enemy to contain. Like NATO in the cold war period, Pakistan was ready to use nuclear weapons to respond to a conventional showdown with India. And the dilemma of India is that it does not know how far it could go to punish Pakistan before Pakistan launched nuclear weapons. To this extent, Pakistan had been fairly unambiguous. Pakistan may be having an edge over India to an extent that its nuclear forces are better prepared [11] and the missile compatible warheads would always have the chances to penetrate the Indian air defence. India of course will be counting on its own nuclear weapons and delivery vehicles like Prithvi, Agni etc., to deter Pakistan. But Pakistani brinkmanship and the hatred of the leadership towards India is well known. And

> "That means New Delhi faces a real dilemma. Any military attack on Pakistan that is big enough to put serious pressure on Islamabad over its support for militants in Kashmir might just be big enough to bring the nuclear option into play. But for India to sit back and allow Pakistan to destabilize Kashmir would be a huge reverse to its plans and position[12] ".

Notes:

[1] Bundy,McGeorge, George F Kennan, Robert S McNamara and Gerard Smith, " Nuclear Weapons and the Atlantic Alliance," *Foreign Affairs*, Spring 1982, Vol 60, No 4 pp753-768.

[2] Columnist Col (Retd) EAS Bokhari, "India-Pakistan-The Nuclear Option".

[3] Group Captain AK Sachdeva, ' Space Age Gladiators" *Knowledge World*, New Delhi 2000 (The Institute for Defence Studies and Analyses).
(Hatf I. The first Pakistani missile programme. It had a range 100Km.
Hatf II is with a range of 300km. The nomenclature was transferred to Chinese M 11s.
Hatf III is a medium rang ballistic missile (MRBM) with a range of 600km.
Hatf IV, range of 500km. Shaheen I with a range of 750km.
Shaheen II, with a range of 2500km.
Ghauri I, This was first named Hatf V and later renamed Ghauri (North Korean collaboration) This is nuclear capable.
Ghauri II, range about 1100km, could be extended to 2300km. First tested on 14 April, 1999.

Ghaznavi. Not much details available. Range is around 2000km. Could have its link with Tapeo Dong.) pp84-91.

4 Group Captain AK Sachdeva, ' Space Age Gladiators" *Knowledge World*, New Delhi 2000 Chinese missiles capability:

ICBM- Liquid Fuelled DF-5/5 A (CSS-5) with a range of 13,000km. Total inventory about 20.

IRBM. DF-3/3A (CSS2 range 2,800km) and the DF-4 (CSS-3, range 4750km). MRBMs DF21/21A (CSS-5), range 1700km.

SRBM- The DF-11 (CSS-7 or CSS-X-7) and CSS-8 are SRBMs with a range of 1000km.

SLBM-Chinese SLBMs include the JL-I (sea going counterpart of DF-21) with a range of 1700 km and the JL–2 under development. JL–2 is expected to possibly contain an MIRV with three or four 90KY warhead). pp68-71.

5 "Nuclearisation of Terrorism," *Indian Defence Review*, Apr-Jun 2002 Vol 17, (2) p88.

6 First strike. A first strike is a surprise attack designed to eliminate the nuclear retaliatory capability of the other side. (An Introduction to Theories of Nuclear Strategy by David G Blair December 1990)

7 Second strike. A second strike is a nuclear retaliation against an enemy that has launched a first strike. This term should be used only in this technical sense. It should not be used to refer to more general defensive moves. (An Introduction to Theories of Nuclear Strategy by David G Blair Dec 1990)

8 Ibid p206.

9 Ibid p213.

10 Ibid p215.

11 Group Captain AK Sachdeva, ' Space Age Gladiators" *Knowledge World*, New Delhi 2000 (The Institute for Defence Studies and Analyses). (The Pakistani missile programme appears to be right from its inception–a reaction to Indian endeavors in the field of developing indigenous missiles. The Indian IGMDP was launched in 1983. Pakistan embarked on its programme soon thereafter, perhaps in 1985.) p94.

12 Huge White, "India V. Pakistan, What Now?" *The Age*, May 30, 2002 < theage.com.au>

Index